THE RELIABILITY FRAMEWORK

Troubleshooting Breakdowns

THE RELIABILITY FRAMEWORK • PART 3

3

Troubleshooting Breakdowns

Systematic Fault Diagnosis for Technicians

MARTIN VAN DEN HOUT

VULCANUS PRESS
FABRICANDO FIT FABER

Colophon

Title:	Troubleshooting Breakdowns: Systematic Fault Diagnosis for Technicians
Series:	The Reliability Framework: Nine Guides to Flawless Asset Performance
Volume:	3
Author:	Martin van den Hout
Photo's:	Depositphotos.com
ISBN:	9789083643144
Edition:	First edition: April 2026

Liability and Disclaimer

The information in this book has been compiled with great care, drawing on 35 years of practical experience in maintenance and reliability engineering. The goal is to provide maintenance professionals with practical tools for improving their maintenance processes.

Neither the author Martin van den Hout, nor Vadeo BV, nor Vulcanus Press can accept liability for the consequences of actions based on this book. Every business situation is unique and requires its own assessment of risks, costs, and benefits.

When implementing the described methods, you should take into account the specific circumstances of your organization, applicable laws and regulations, safety requirements, and manufacturer guidelines. When in doubt, consult specialists and make responsible decisions based on thorough risk analysis.

This book is intended as a practical guide and does not replace customized professional advice, adequate employee training, or sound engineering judgment.

Foreword

Anyone who works daily to keep their assets reliable knows how exceptional that profession is. Behind every running pump, every production line that keeps operating, and every network that keeps functioning, stands a skilled professional with knowledge, dedication, and an unstoppable curiosity for improvement. The maintenance professional is the silent force that makes technology perform, monitors processes, and moves organizations forward — often without anyone noticing, until something goes wrong.

Yet the world of maintenance is changing faster than ever. From TPM to Lean, from Asset Management to Reliability Engineering – the terminology shifts, but the core remains the same: reliability starts with people who know what they're doing, and who dare to learn, share, and improve.

This book is an invitation to you – the professional who looks beyond repairing what is broken. It is a compass for those who view maintenance not as a cost, but as the engine of continuity, safety, and value creation. Whether you work in industry, the public sector, or at a high-tech company, the principles shared here are universal: structured thinking, craftsmanship, and working together toward reliability.

Let this book inspire you to look beyond today. See every maintenance moment as an opportunity to learn, every problem as a reason to improve, and every asset as a system that can do better — because you want to be better.

My admiration goes out to Martin van den Hout, author and knowledge custodian with decades of experience in the maintenance and Asset Management field.

His sharp insights, thorough knowledge, and passion for sharing expertise have made an invaluable contribution to this work. Thanks to his ability to make complex subject matter clear and accessible, his contributions form a valuable bridge between theory and practice.

Because, ultimately, maintenance is not just a profession: It is a mindset!

Ellen den Broeder-Ooijevaar
Manager NVDO
(Dutch National Maintenance Society)

WARNING!

In this book you will read about finding the root causes of failures in technical systems and resolving those failures. Never attempt to do this if you are not qualified or do not have a clear understanding of how the system operates. Working on technical systems can be dangerous and may result in serious injury or even death. It is essential to comply with all applicable laws and regulations, both from your company and from governmental authorities.

Table of Contents

Introduction

This book is part of a series that answers the question: "What does an organization need to do to make its technical installations perform as well as possible?" I asked myself this question 35 years ago, and it has kept me constantly occupied ever since.
I was working at the time in the Maintenance & Engineering department of a Japanese manufacturing company in the Netherlands. Back then, the term "Total Productive Maintenance" (TPM) was on the rise, and more and more people were coming to realize that simply setting up a good maintenance schedule is not enough to keep a production plant running stably.

Within TPM, the emphasis was strongly on involving production operators and improvements within a production department. I realized fairly quickly that these were major steps forward, but still not enough.
There were also many questions within our maintenance department that TPM didn't answer. Questions like: "Which spare parts do we need and how many?" or "What is the best way to organize the large flow of work within a maintenance department, ranging from jobs that take a minute to projects requiring thousands of man-hours?"

Over the past 35 years, our field has matured. TPM was incorporated into Lean Manufacturing, and from (semi-)governmental bodies came increasing attention to technical management, under the name asset management. This ultimately led to the publication of the ISO 55000 standards series. For the reliable operation of a production plant or other installation, the term "Reliability" emerged as a replacement for the term "Maintenance." After all, reliability is the goal, and maintenance is merely one of the means.
I learned that all these movements strongly resembled one another and that the differences were primarily differences in emphasis, caused by corporate culture and, for example, by factors such as how many operators are directly involved in production. In the chemical industry their role is different than in the automotive industry, and in government organizations this role is different again. Ultimately, many organizations have

combined elements from these approaches in whatever way best suited their specific situation.
Yet I learned that what applies to a production plant also applies to a network, your car, or a power plant. It makes no difference.

It always bothered me that there were only two kinds of books about this field.
The first type describes the entire field, but only describes 'what' you should organize. They are therefore very broad, but never go into depth on any subject.

The second type focuses on one aspect, such as work flow management or setting up preventive maintenance, and goes into great depth on how to apply that one aspect. These books have a number of things in common. They lack coherence with one another, there are very few good ones, they are usually written in foreign languages, and they are always very thick. They are also often written from the perspective of "the ideal world," in which you have an infinite number of engineers or technicians at your disposal and the most fantastic computer programs.

How This Book Is Organized

This book deliberately follows the natural sequence in which you address a failure — from notification to follow-up care. You can think of it as a journey through five crucial steps, where each chapter builds on the previous one and systematically guides you toward the solution.

Understanding why working systematically pays off (Chapters 1-3) begins with a confronting case: twelve hours of chaos caused by a loose plug.

Chapter 1 shows you what can go wrong without a systematic approach.
Chapter 2 introduces the 5-step methodology that enables failure technicians worldwide to work 50% faster.

Chapter 3 covers the theoretical foundations — how technical systems actually fail. You will learn about the 5M's (Man, Material, Method, Medium, Machine), discover that 68% of failures occur shortly after installation or maintenance, and understand why changes are the primary cause of failures.

From notification to diagnosis (Chapters 4–7) is where you learn to ask the right questions and gather information.
Chapter 4 covers failure reporting and effective communication under pressure. The 7WH method (What, Why, How, Where, When, Which, Who) becomes your compass. You learn to deal with stressed operators, management under time pressure, and the prevention of conflicts during a crisis.

In chapter 5 you go on-site: the initial reconnaissance with systematic observation, verification of information, and documentation of the original situation. Here it becomes clear why "looking before you act" saves time.

Chapter 6 covers the formulation of possible causes: from consulting failure history to modern digital tools such as smartphone diagnostics, AR, and AI support. The 5M methodology is applied practically here for structured root cause analysis.

Chapter 7 covers modern tools you can use, from your phone to augmented reality and from digital documentation to real-time monitoring. You will read about these developments and how you can integrate them into the systematic approach to troubleshooting failures.

From Analysis to Solution (Chapters 8–10) concretely illustrates the difference between trial-and-error and systematic work.

Chapter 8 teaches you to ask the core question: "Does this cause explain all the symptoms?" You learn to recognize tunnel vision, avoid thinking errors, and systematically eliminate causes without unnecessary repairs.

Chapter 9 continues with root causes through the 5-why rule and vertical thinking. Here you don't stop at the symptom but find the true source of the problem.

In chapter 10 you prepare the repair with a focus on safety. LMRA (Last Minute Risk Analysis) and LOTOTO procedures receive the attention they deserve — after all, 75% of industrial accidents occur during maintenance. The warning is clear: "hurry up" and "real quick" are the most dangerous words in failure repair work.

Testing and learning (Chapters 11–13) ensures that your work has a lasting impact.
Chapter 11 describes phased testing and systematically dealing with failed repairs — because not every repair succeeds the first time.

Chapter 12 covers effective failure reporting as the foundation for continuous improvement and knowledge sharing within the organization. Good documentation prevents the same failure from costing hours of troubleshooting again later.

Chapter 13 closes with follow-up care: from horizontal thinking (could other assets have the same problem?) to preventive implementation that prevents future failures.

Prepared (Chapter 14) looks ahead at what you as a maintenance technician and your organization can do to be better prepared. From personal toolbox to modern digital systems.
Each chapter closes with a natural transition to the next and includes practical case studies of cooling compressor C-15 that bring the theory to life. The appendices offer directly applicable templates: the guide form for the 5-step methodology and the complete 7WH questionnaire. The central theme is clear: with a systematic 5-step approach you resolve failures much faster, avoid replacing unnecessary parts, and work more safely. The question is not whether you should learn this method, but whether you can afford not to.

The Complete Series

This series consists of nine books, which together provide the answer to the previously mentioned question: "What should an organization do to make its technical installations perform as well as possible?" The books cover one topic each in a systematic, step-by-step approach. Each book connects to the other books in the series. This way, you can implement a coherent system within your company to allow your assets to perform optimally.

This is the third book in this series. The series consists of the following books:

Part 1: Maintenance Management

A comprehensive overview of the entire methodology. A general description of the topics covered in the other eight parts, plus several additional subjects.
ISBN 9789083643120

Part 2: Planning and Scheduling in Maintenance

Organization of the workflow of the maintenance department; including work requests, gate-keeping, work preparation, scheduling, supervision, safety, administration, improvement, and archiving.
ISBN 9789083643137

Part 3: Troubleshooting Breakdowns

The resolution of an individual failure, by a maintenance technician or an operator. Notification, finding the cause, finding a solution, repair, failure reporting, continuous improvement.
ISBN 9789083643144

Part 4: Root Cause Analysis in Maintenance

Reducing the number of failures that occur annually. Registration, statistical analysis, prioritization, root cause analysis, and implementation of solutions.
ISBN 9789083643151

Part 5: Preventive Maintenance

Setting up a preventive maintenance schedule for an entire organization, taking into account the risks and criticality of the individual assets. Translating management requirements for the entire organization into technical requirements for the assets. Mapping possible failure modes, causes, and possible maintenance actions. Economic approaches to determine which maintenance is worthwhile. Methodologies such as RCM, Industrial RCM, and pragmatic approaches.
ISBN 9789083643168

Part 6: Spare Parts Management

What types of spare parts are there? Is it worthwhile to keep them in stock? How many of each type? Organization of the warehouse and key considerations for common types of components.
ISBN 9789083643175

Part 7: Precision Maintenance

Many assets fail prematurely because they have not been assembled with sufficient precision. A technical handbook covering practical topics such as bolt tightening, fits, aligning rotating equipment, assembly with attention to balance, flange assembly, and so forth.
ISBN 9789083643182

Part 8: Engineering for Reliability

80 percent of maintenance costs and technical reliability is determined during design and construction. Project management, Early Equipment Management, involvement of operators and technicians, reliability analyses during design, FAT and SAT, Commissioning, Maintainability.
ISBN 9789083643199

Part 9: Autonomous Maintenance

Which maintenance tasks can you have operators perform, why, and how do you organize that?
ISBN 9789083706306

For a current overview of all books, additional information, and ordering options, visit:

vadeo.nl/books.html

1. Twelve Hours of Chaos

It was 3:47 in the morning when the phone rang at ServiceTech Inc. "Complete failure at FreshFood Industries," the voice on the other end gasped. "Our cold chain has gone down completely. We're losing $20,000 per hour in perishable products. You need to get here NOW!"
James Campbell, a failure technician with ten years of experience, threw on his clothes. Upon arrival, he saw a chaos of flashing alarm lights and stressed operators. "The main motor of cooling compressor K-1 has completely stopped working," the production manager shouted. "Just replace it, we have spare parts!"

James felt the pressure. Twenty thousand dollars per hour. He glanced briefly at the motor — no smell of burning, no visible damage — but the pressure was immense. "Okay, put a new motor on it," he decided without further analysis.
Two hours later, with a brand-new $15,000 motor installed, there was still no cooling. Worse: the new motor was making strange noises. "Maybe the control circuit," James mumbled, sweat on his forehead. He began replacing relays at random.

At 7:00, colleague Jack Wheeler arrived for the day shift. "Electrical failure," Jack concluded immediately after James's account. "Must be in the control panel." Jack dove into the main control panel and began systematically replacing components. After 3 hours, he had replaced eight different parts, but the failure persisted.
Meanwhile, conflicts arose. The production manager shouted at the operators: "Did you do mess with the settings again?" The operators pointed to maintenance: "Someone worked on it just yesterday!" James and Jack were now working against each other, each convinced of their own theory.

By 11:30 the situation had escalated. Two technicians had been working without informing each other. Two systems had now been modified, nobody knew what the original condition had been anymore, and parts were scattered throughout the entire machine room. The stress was so high that Jack began checking energized components without safety measures.

The climax came when James tried to reconnect the "new" motor while Jack was still taking measurements in the switchgear cabinet. A short circuit caused an explosion in the control panel. Now there were two faulty motors and a burned-out controller. Shaken, they began installing the last spare parts they had left. They decided to bring in senior technician Keith Baker, who was actually on a day off.

At 2:15 PM — more than ten hours after the original report — Keith arrived. Instead of getting straight to work, he did something different: he began systematically asking the operators questions. "When exactly did it start? What did you hear? Which alarms came first?"
After just five minutes, Keith asked: "Who worked on the system yesterday?" A junior electrician hesitantly raised his hand. "I replaced a sensor in cooling conveyor T-12..."

Keith walked over to the conveyor belt. There, half hidden behind a cable duct, sat a loose connector. A simple plug connection that had come undone during the sensor work. The connector provided the temperature measurement that controls the cooling compressor. Without a signal, the system assumed no cooling was needed.
Five minutes later, after securing the connector, the cooling compressor hummed back to life. The temperature began to drop. After twelve hours of chaos, the problem had been solved with a thirty-second hand movement.

The damage was enormous: a perfectly good motor worth $15,000 destroyed, a control panel worth $8,000 burned out, eight useless spare parts consumed, and more importantly — nearly $240,000 worth of spoiled food. Worst of all: a near-accident caused by unsafe work practices under extreme pressure.
"How is this possible?" the director of FreshFood Industries asked. "A loose plug and you search for twelve hours?"

Keith shook his head. "They made the classic mistakes: drawing conclusions without analysis, no systematic approach, and working under pressure without safety measures. With a structured method, this would have been resolved in half an hour."

That evening, James sat at home, exhausted and frustrated. He realized that his ten years of experience had taught him how to repair machines, but not how to systematically analyze a problem. He had reacted instead of acted, guessed instead of reasoned.
The bill came due: ServiceTech had to compensate for the damage to the installations. Their reputation had taken a hit. James realized there had to be a better way.

That better way exists. It is no secret, no magic, no expensive equipment. It is a systematic 5-step methodology that makes failure technicians around the world 50% faster than a trial-and-error approach. A method that prevents you from replacing expensive parts that are not defective, that helps you work safely under pressure, and that ensures you find the actual cause instead of merely treating the symptoms.

Chapter 2 covers the overview of the entire methodology.
The question is not whether you should learn this method. The question is: can you afford not to learn it?

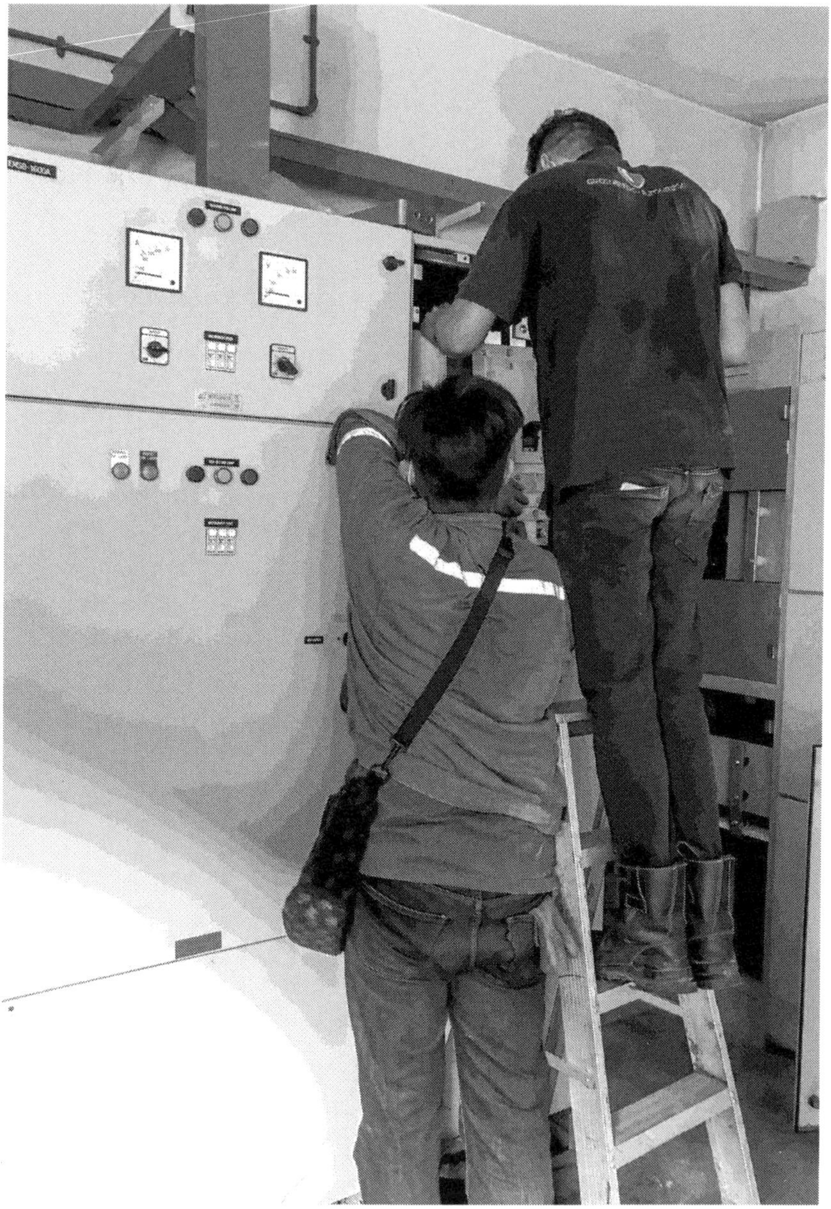

2. The Main Steps of the Approach

In this book, the word ‘failure’ is used for any situation in which an asset does not meet its requirements. So it may be that the equipment is not working at all, but it could also be that it makes too much noise or is leaking. A flat tire is a failure, even if you could still ride. In some cases, one person may consider a situation a failure, while another is perfectly satisfied with the way an asset is functioning. It depends on what requirements you place on an asset. These are often not concrete or not fully specified. A failure can also be called a defect or a fault. In this book we use all these terms interchangeably, meaning that the asset is not functioning properly.

Experienced breakdown technicians often resolve failures much faster than their inexperienced colleagues, not because of intelligence or superior technical knowledge, but because of their mastery of a working method for tackling failures effectively.
If you are troubleshooting a failure, you are a breakdown technician. That is the term used in this book, whether you are solving this failure as a professional at work or at home on your own equipment, or whether the problem occurs in a mechanical system, electrical system, electronics, chemical process, or software. The logic of the approach is the same.

This book explains not only how to find the cause of a failure, but also how to resolve it. The book follows the step-by-step methodology as closely as possible. To resolve a failure quickly and effectively, you must first find the cause, since repairing a defective asset without knowing the cause is usually impossible.
Suppose the temperature of an industrial furnace is too low. You could of course try to set the temperature a bit higher. That might resolve the symptom of the failure. However, if the temperature is lower than it should be at its current setting, then something more is going on. Perhaps there is something wrong with a controller, a burner, insulation, or there is some other cause. If you don't eliminate it, it usually gets worse and the problem returns after a while. At some point, it is no longer possible to set

the temperature of the oven any higher. With many failures, the option to adjust a setting isn't there at all. You therefore need to find the root cause in order to truly resolve the failure. Once the cause of the failure has been identified, it is usually fairly clear what needs to be done to fix it.

Sometimes the cause is immediately obvious and you can begin the actual repair right away. In many cases, however, you first need to investigate what the cause is. If you don't know the cause, you can at best treat the symptom. The danger of this is that the failure keeps recurring or even gets worse. Finding causes requires a great deal of knowledge and skill.

Naturally, a maintenance technician who knows the asset in which the failure occurs has a significant advantage. If you don't know how a car works, it will be very difficult for you to repair it. In addition, general technical knowledge is required, for example in the field of mechanical engineering or electrical engineering. But what is especially needed to properly resolve a failure is an analytical method.

2.1. The 5-Step Methodology

To resolve a failure, you must systematically work through a number of steps. In this book you can read about these steps and how to apply them. The steps are:

1. Describe the failure clearly and concretely Very often it turns out that people spend unnecessarily long amounts of time solving a failure, because they don't know exactly what the symptoms of the failure are: what is happening exactly, how often, where, how it is detected, and so on. This symptom analysis sometimes takes only a few minutes, but saves you hours later. So before you can look for the cause, you need to know as much as possible about the failure itself and its symptoms. To put together a good description of the failure, you need to ask the right questions of the person who reported the failure to you. You also need to ask these questions in the right way, to ensure that your conversation goes smoothly. Usually you also need to inspect the failing system — look, listen, smell, or take measurements — to find the cause.

2. Draw up a list of possible causes You will read about various methods to find the cause.

3. Verify that the possible cause actually produced the problem The cause must explain all the symptoms — not just, for example, that a machine damages the product, but also that it happens at certain times and not at others, or that it damages the product in only one location. This is a very important step. If a potential cause does not explain all the symptoms, it cannot be the true cause of your problem. Some people even say that if you truly know all the symptoms, there can be only one cause of your problem. All others can be eliminated.

4. The Repair In some cases, there is only one possible cause that explains all the symptoms. Often, however, there are several. In that case, you must choose which cause to address first. You can do this by thoroughly resolving the failure and eliminating the cause, but in the real world it is sometimes better to opt for a temporary solution or to limit the consequences. If a pipe is leaking, you can repair the pipe or place a collection tray underneath it. In this phase, the most important thing is to think about how you can safely resolve the failure. Unfortunately, troubleshooting failures is one of the most dangerous activities in modern industry — 75% of all accidents occur during maintenance activities.

5. Aftercare and Follow-Up Once you have resolved the failure, you are not done yet. You still need to:

- Think about the underlying causes and further consequences of the failure
- Test whether the system is working properly again
- Write a Failure Report
- Think about how to ensure this failure never occurs again

For comprehensive Root Cause Analysis (RCA) of recurring failures, see book 4 in this series: Root Cause Analysis in Maintenance.

The Importance of the Correct Sequence

The order of the five steps above is very important. In many cases, when a maintenance technician has had to search for a long time for the cause, it turns out afterward that he drew conclusions too quickly and started replacing or "repairing" parts that were perfectly fine. If he had asked a few simple questions, looked carefully, and thought things through, he could have concluded within minutes that the cause he had in mind could not possibly be the cause. He was trying to find a cause while he did not yet know the problem.

It takes less time to ask a few questions, observe, and think than to try to solve the problem in an unstructured way. It is also much safer for you and your equipment to not work on it any more than necessary.
By asking open-ended questions first, you can quickly rule out possible causes. Then you ask closed-ended questions to verify your theory about the failure. Only after you have taken measurements and established the cause do you begin the actual repair. If you follow this sequence, you will usually find during testing that the system is working properly again.

If you don't work in the correct order, you can waste a lot of time replacing parts that weren't defective. When you test the asset after this work, you discover it still doesn't work. You may have even damaged components that were working fine when you started!

3. How Do Technical Systems Fail?

To become proficient at troubleshooting failures, you first need to understand how they are generally caused. There are many misconceptions about the causes of failures. You might think that assets fail because parts wear out and then break down. That is true for only a small portion of failures.

Another common misconception is that failures only occur in assets that are old and worn out. This is incorrect. Some types of failures actually occur more frequently in new assets.

3.1. Categories of Causes

Often the cause of a failure is not technical. It is not always the case that something has worn out, broken down, rusted through, become contaminated, or whatever other technical causes you might think of. When you need to resolve a failure, it is helpful to keep the following five groups of possible causes in mind:

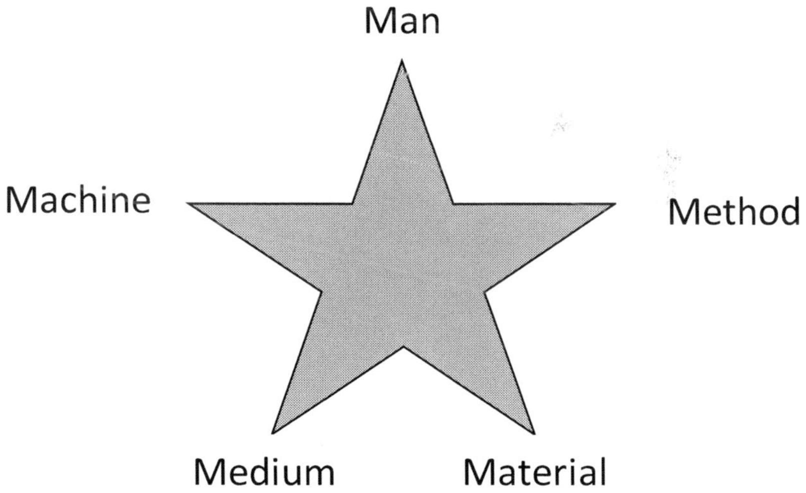

Figure 3-1 The five M's

You will probably notice that these five words all begin with the letter 'M'. This makes them easier to remember. In the following sections you can read what we mean by these five words

Man

At many companies, more than 50% of failures are caused by human error, such as operating errors or assembly errors. Failures often occur shortly after maintenance has been performed on an asset. No matter how experienced people are and how well they normally do their work: everyone makes a mistake sometimes or overlooks something. Never assume that everyone has done their job correctly. A good breakdown technician tries to find out exactly how people have been working with the asset, without accusing them of making mistakes. That can be quite a sensitive issue. Everyone makes a mistake from time to time.

Material

The use of incorrect materials in a machine can also lead to problems. Materials include, for example, raw materials, packaging materials, or tools. Feeding incorrect raw materials into a system can cause all kinds of issues. If a sewing machine suddenly has to sew with a thicker thread, this can lead to problems. If a product with a higher viscosity has to be mixed in a mixing tank, the agitator may fail. The use of different degreasers can strongly affect the results of spraying in a paint shop.
By "wrong" materials we mean materials that the asset cannot process properly. These can also be materials that the asset should be able to process according to the manual or its design. In some cases, the asset simply needs slightly different settings and everything works fine again. We use the word "wrong" here because with the current settings it does lead to a failure.

Method

'Method' in this case refers to all specifications, setup lists, operating instructions, and similar documents that exist for the asset. The user can do exactly what these documents prescribe, but if they contain errors, failures still occur.

Errors in the 'method' can be divided into two groups:
In the first group, the knowledge is present within the organization to write a good method, but it has not reached the user. This can happen, for example, because the user is working with an outdated version of a setup list or because it has not yet been updated.

In other cases, nobody knows exactly how the equipment should be operated. This often arises when companies want to keep increasing the output of their assets. Suppose a machine is designed for ten thousand products per hour. If demand for the product increases, the company may want to run the machine faster and produce eleven thousand products per hour. If this is possible with the existing machine, it is much cheaper than investing in a new machine. If demand rises further and the company keeps setting the machine faster, a point will come where this is no longer possible. Perhaps we should say: barely possible. The engineers may then set the machine to thirteen thousand products per hour. If that works for a while, they make it the standard setting. If some time later one variable changes slightly, the machine can fail, simply because it has become very sensitive to fluctuations in temperature, humidity, viscosity of the raw material, or vibration.

Medium

'Medium' is another word for surroundings. Failures can also occur due to conditions in the environment of the machine. Think of vibrations, moisture, or high temperatures. Often it is a change in conditions that initiates a failure. A rapid temperature change, for example, can cause failures in electronics.

Machine

Of course, a failure can also occur due to technical failure of a part of the asset. By 'Machine' we mean all technical causes. These are caused by underlying mechanisms, such as:

- Wear
- Friction/Abrasion
- Corrosion

- Effect of Electromagnetic Fields on Plastics
- Fatigue
- Overload
- Loose Components
- Built-in components of poor quality

3.2. When Do Failures Occur?

Some causes typically occur with aging assets, but a number of others occur precisely with newer ones.

If you look at all the ways assets can fail due to causes in the 'Machine' category, it turns out that only 11% of failures show an increasing probability of failure as the component gets older.

It is therefore not the case that only old components fail. In fact, in 68% of failure modes there is an increased probability of failure during the initial period following installation or maintenance.

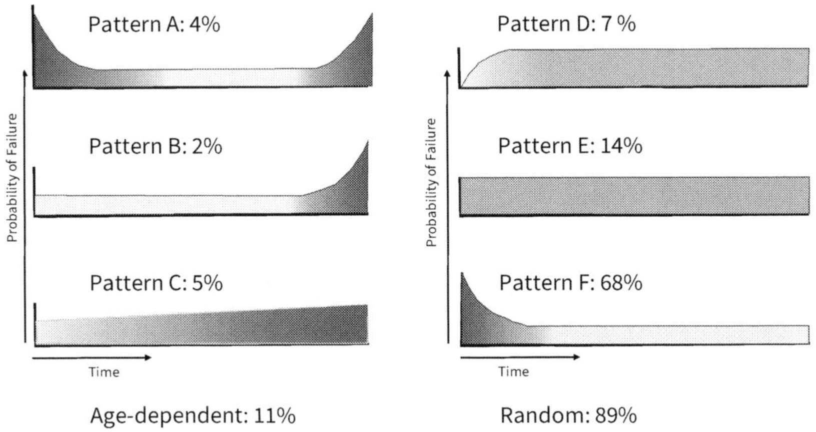

Figure 3-2 The six failure patterns

On the vertical axis in Figure 3-2, you can see the probability that a given failure mode will occur in the upcoming period; on the horizontal axis, the passage of time.

One failure mode is, for example, a bolt rusting through, another is a bolt coming loose, and yet another is a bolt suddenly breaking due to extreme overload.

A new bolt will not yet have rusted through, but over time, under the wrong conditions, the likelihood of corrosion increases. This failure mode falls under failure pattern B. Fracture due to fatigue of the bolt falls under pattern C. Fracture due to extreme overload can happen at any time, so this falls under failure pattern E. The loosening of a bolt occurs primarily shortly after the bolt has been tightened. If it turns out to be properly secured, there is little chance that it will come loose. This failure mode therefore falls under failure pattern F.

Figure 3-2 shows that 68% of failure modes conform to failure pattern F. After a certain run-in period, failure patterns D, E, and F are all random. Together these account for 89% of all failure modes. Note: a failure mode is different from a failure. A failure mode is a way in which a failure can occur. A failure is one instance of a bolt coming loose or breaking. A failure mode is the way in which it happens. So ten failures that occur because bolts come loose all have the same failure mode.

In the figure you can also see that some failures typically occur more often when an asset is old, and some other failures occur more often when an asset is new. It is a (common) mistake to say: 'this part is new, it cannot cause any problems.'

In many books about failures and failure modes, you will find the 'bathtub curve.' This is the graph of failure probability over time for an asset as a whole. It broadly resembles failure pattern A, but is not strictly the same. It is the graph that results when you add up all the failure patterns for an asset.

Changes as a Reason for Failure

The vast majority of failures occur as a result of an 'event.' The events can occur in various ways, sometimes just before the moment of failure, but sometimes much earlier. An event can be anything. Perhaps the asset is under a heavier load than normal. It could also be that maintenance has been performed and some components have been replaced. Perhaps the asset got wet or was filled with new raw materials. Figure 3-3 illustrates this.

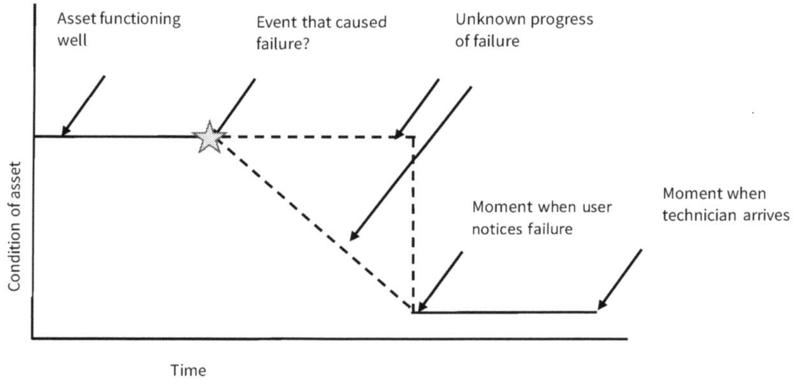

Figure 3-3 Progress of a failure prior to notification

On the vertical axis of this figure you can see the 'quality' of the asset. "Quality" in this case means a property that you want the asset to have. It can refer to things like speed, force, surface roughness, temperature, pressure, or any other possible property of the asset. On the horizontal axis you can see the time that passes. We assume that the asset once had sufficient quality. It was therefore working properly. At a certain point, the breakdown technician is called to the asset, because the quality is no longer acceptable. At some point in the past, something happened that caused the quality of the asset to decline. We refer to this in the figure as an 'event'. The dashed lines in the figure indicate the ways in which this may have occurred. Perhaps the asset failed immediately after the event and is no longer functioning at all. It may also be that it gradually performed worse and has now stopped working entirely. However, it may also be that it gradually performed worse and is still working, but slightly below the quality limit that we consider acceptable.

The way in which the asset 'deteriorated' reveals a great deal about the cause of the failure. A gradual decline, to just below a rejection threshold, usually occurs due to a gradual process such as contamination or wear.
If an asset has always worked well, it usually doesn't just fail out of nowhere. A number of components will wear down slowly. They become "weaker" and cannot handle as much load as when they were new. This can ultimately lead to failure of the component. Figure 3-4 shows this schematically.
In this figure, the component fails when the 'strength' of the asset drops below the level of the load on that component. Take for example a steel structure that slowly rusts through or a V-belt that gradually wears down. Even if the load is constant, they will eventually fail.

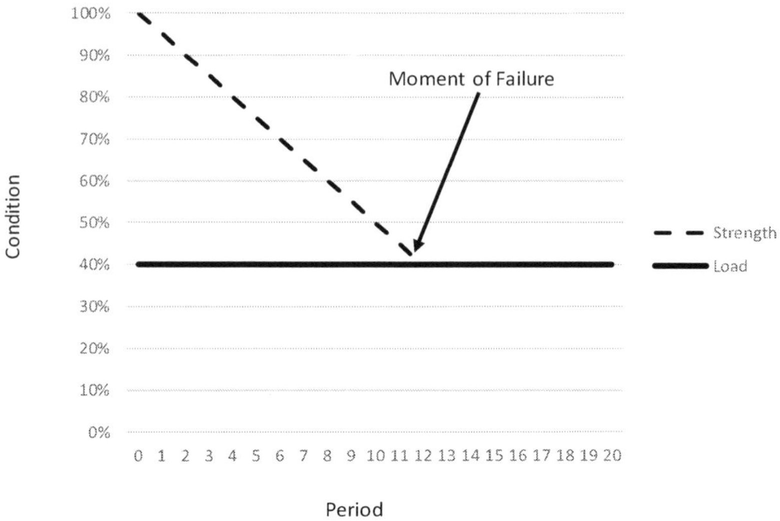

Figure 3-4 Failure under a constant load

Weak components, however, often fail when they are subjected to greater loads than usual.

This can be caused by a heavier or more viscous raw material, high ambient temperature, extra weight being placed on the steel structure, or for example by starting up the asset. You can see this scenario in Figure 3-5.

An example of this is a family car that is used throughout the year solely for commuting. This is a daily distance of 20 kilometers one way, with only the driver in the car in reasonably cool weather.

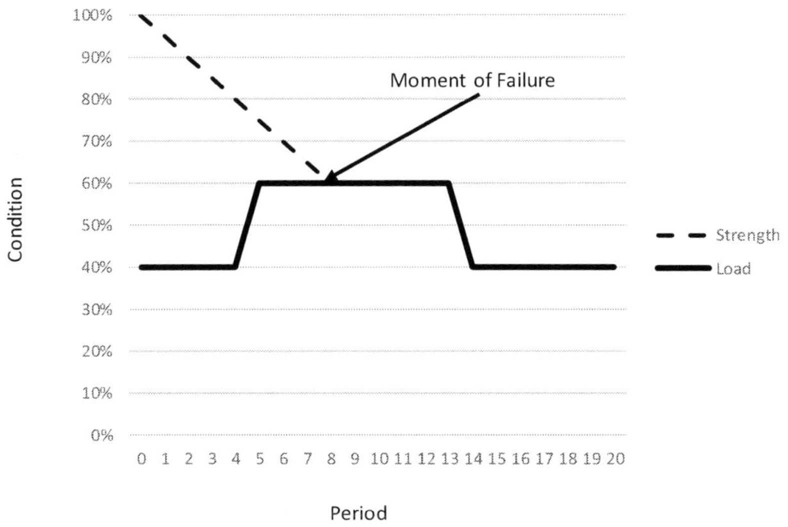

Figure 3-5 Failure under a changing load

If this car suddenly has to travel a thousand kilometers to a vacation destination in the summer, with four people inside, a trailer attached, and a roof box on top, in the heat, then a number of components will be placed under much heavier loads than during daily commutes. It is precisely in these kinds of exceptional situations that more failures occur.
In this schematic graph, the component fails in period eight instead of in period twelve.

Components rarely fail when the asset is simply loaded as usual. If you know whether the asset is processing different raw materials than normal, or that it has just been started up, this gives you insight into the type of components that might fail. If you know what has happened to the asset in the last few hours or days, you can deduce a great deal about the possible cause of the failure.
Now that you understand how failures truly occur — not just through wear but through many different factors — you can apply this knowledge directly when gathering failure information. The 5M's give you a systematic framework for asking targeted questions. In Chapter 4, you will learn how to collect the first crucial information from the person reporting the failure.

4. The Failure Report and Questioning Techniques

In this book, we cover troubleshooting a failure as much as possible in the order in which you will also work in practice. Usually it starts with someone reporting the failure to you. Failures can be reported to you in all kinds of ways, for example through an Enterprise Asset Management (EAM) system. Maybe someone comes to you in person or calls you.

Modern Enterprise Asset Management (EAM) systems enable digital failure notifications with automatic asset identification and failure history. These systems support the 5-step methodology through better documentation and faster access to relevant information, but do not replace personal contact with operators for collecting crucial details.

In this chapter, you will read a number of questions you could ask. It depends on the situation whether you ask all these questions by phone or whether you decide, after a few questions, to go to the location of the failure and ask the questions there. A maintenance technician who has to travel before reaching the location of the failure should ask as many questions as possible by phone. Perhaps he can already provide enough information by phone to allow the people on-site to resolve the failure themselves. If he does go to the failure, he can at least bring all the right parts and tools. If someone is close to the location of the failure, he can often do better by going there immediately. He can then see with his own eyes what is going on. How serious is that "major leak" really? What does that "scraping noise" actually sound like?

Before taking any action, you must first talk to the person who reported the failure. After all, it is his problem that you are trying to solve.
You already read that a system fails or has a failure when it does not meet its requirements. These requirements, however, are often not clearly specified. In that case, the requirements are not objectively measurable, but subjective. This means that the person who reports the failure to you

thinks there is a problem, while someone else might find the situation acceptable or perfectly normal.
In office buildings with an ideal temperature, for example, there are always people who find it too cold and people who find it too warm, at the same moment. Another example is a system that has been leaking a little water for months. At a certain point, someone comes along who finds it unacceptable and reports the leak. All the other people who saw the leak apparently didn't consider it serious enough to report. This person did. So the fact that you are called in today does not mean that the leak started today.

In general, you can say that a system has a failure when it functions in some way not as well as normal, and then so much worse that someone has a problem with it.
It therefore depends on the person who notices the failure whether he thinks it is a real failure. He can tell you exactly what he is having problems with and what the failure looks like in his eyes. In any case, make sure it is clear to you what the person reporting the failure is having problems with. He can probably also bring you much closer to the solution. As a maintenance technician, you often arrive at a piece of equipment that "isn't running." In practice, this usually means it has stopped working altogether. It can then be difficult to find a starting point. The person who reported the failure may have been present when the problem started. He can tell you exactly what happened.

Case Study: Urgent Report – Cooling Compressor C-15

At 2:30 PM, supervisor Lisa from MediCare Distribution calls: "Our refrigeration compressor C-15 is making a strange noise and the temperature is rising. We have $50,000 worth of insulin in that cold storage unit that needs to be kept at 8°C. It's now at 7°C and still climbing!"
This notification combines technical symptoms (noise, temperature rise) with business urgency (valuable medications). Lisa provides concrete measurements, but is clearly stressed. Before heading to the location, you need to gather more specific information about the symptoms to prepare your diagnosis.

4.1. Question Techniques and Communication Under Pressure

Before we describe what questions you can ask, we first address how you ask those questions and how you communicate effectively, both in normal situations and under the stress of failure situations. By asking questions correctly, you are more likely to:

- Getting the Right Information
- Getting it as quickly as possible
- Maintain a good relationship with the person you are talking to.

4.2. Fundamentals of Effective Communication

People often don't provide the information you need to solve a failure. Often this is simply because they don't realize what that information is. An operator often doesn't realize that he knows much more about the process and the operation of a machine than you as a breakdown technician. It can also be that people have too strong a tendency to take action (or to rush the breakdown technician). You then get comments like: "Stop asking so many questions, just replace that valve so we can start up again."

Never Give a Negative Judgment about Answers

If you attack or judge someone when they honestly answer a question, they will not be as honest when answering the next question. So don't say: "What were you thinking doing that?" or "You should never have done that!" It is best to remain as neutral as possible. Remember that more than 50% of failures are caused by human error. We all make mistakes sometimes.

Adapt to the Other Person

Conversations flow much more smoothly when you can put the other person at ease. Adapt your speaking pace, word choice, and body posture to those of your conversation partner. If someone speaks slowly, speak a little more slowly yourself. If someone rushes into your office and talks very fast, however, it is often better to start talking quickly yourself as well. After a sentence or two, you can slow down to a calmer way of speaking.

Use Plain Language

Many specialists unconsciously use technical jargon. People who do not work in that field do not understand it and quickly lose their attention. Feel free to use jargon with a fellow technician, but phrase questions for others in language they can understand.

Listen Carefully

Good listening is one of the most demanding things there is and encompasses two aspects:
Truly paying attention to what the other person is saying
Let the other person know this as well by looking at him, summarizing key points together, and giving confirmation by nodding.

4.3. Communicating Under Pressure

Failure situations always bring extra stress. Production is at a standstill, people are frustrated, and everyone wants the problem resolved as quickly as possible. In this tense atmosphere, you as a maintenance technician still need to calmly ask your questions and gather the right information.

Dealing with Stressed Operators

Operators are under enormous pressure when their machines fail. This stress manifests itself in different ways: some operators become quiet and withdrawn, others start talking more, and still others become defensive and irritable.

With Quiet Operators: Start with simple, concrete questions and build up slowly. Show understanding: "I understand this is causing stress. The faster we have the right information, the faster I can resolve it."

For Operators who Talk too Much: Help by providing structure: "Those are all important points. Let's go through them one by one so we don't miss anything." Use the 7WH method that you'll read about later in this chapter as a guide.

Handling Management Pressure

Managers feel the financial pain of downtime directly and often want to know: "How long is this going to take?" This pressure can make it tempting to quickly guess at the cause, but this usually leads to even more delay. "I'm going to systematically map out the symptoms now. This will take fifteen minutes, but it prevents me from spending hours replacing parts that aren't defective." A systematic approach is much faster than trial and error.

Preventing and Resolving Conflicts

During failures, conflicts sometimes arise where operators point to maintenance and vice versa. Stay neutral and focus on facts instead of blame. Use phrases like: "Let's look at what exactly happened happened" instead of "who did this." Emphasize that your goal is to solve the problem, not to find someone to blame.
When people become defensive, acknowledge their expertise: "You know this machine best. What did you notice before the problem started?"

Communicating Technical Information Clearly

Use analogies that people can understand. Instead of "the PID controller oscillates due to incorrect tuning" you can say "it's like a shower that alternates between too hot and too cold, because the faucet is too sensitive is set."

Adapt your Explanation to your Audience:

For operators: more detail on the impact on daily work
For managers: focus on impact and resolution time
For technical colleagues: specific technical terms

Practical Tips for Stressful Conversations

- Speak more calmly than normal - this has a calming influence
- Take the time to deliberately think before you answer.
- Move away from a noisy environment if possible before important conversations
- Repeat what you hear in your own words: "If I understand you correctly, you're saying that..."
- Show empathy: "I can see this is causing a lot of stress"

In stressful situations, people make communication errors more easily, so extra verification of information is crucial. Now that you have read how to communicate with operators to efficiently obtain the right information, the next section covers which questions are best to ask.

4.4. Open-Ended Questions

Do Not Start with Closed-Ended Questions

Closed-ended questions are questions that can only be answered with 'yes' or 'no'. Closed-ended questions are often also leading questions. Leading questions are questions in which you already suggest an answer to your conversation partner. When you ask them, you are not open to all possible answers, but instead impose your own ideas on your conversation partner. So do not ask questions like: "Is it slowing down?", "Is the product warmer?" or "Is the belt misaligned?"

Suggestive questions sometimes lead to conflicts. The question: "Did you run the agitator too fast?", will cause irritation among the operators. You are essentially accusing them of vandalism. The person you are asking the question to will feel (rightfully) attacked. He will be less willing to cooperate with you.

In a number of other cases, suggestive questions may not lead to conflicts, but they do lead to incorrect answers. Be very careful with questions such as:
"Did you hear a rattling noise?" The person the question is being noise.' He now tends to answer affirmatively, even though he heard something he himself would have described differently than as a 'rattling sound'.
At a slightly later stage, when you have narrowed down the number of possible causes to just a few, suggestive questions or other closed-ended questions can be useful.

Start by Asking Open-Ended Questions

A very practical way to avoid putting words in your conversation partner's mouth or accusing him of something is to ask open-ended questions. Open-ended questions usually begin with interrogative pronouns:

- What?
- Why?
- How?
- Where?
- When?
- Which ones?
- Who?

These questions are called 'open' because they don't steer the answer in any particular direction. A good open-ended question to start with is: 'What's going on?' or 'What's your problem?' The person giving the answer then has to say something more than just 'yes' or 'no'.
Open-ended questions force the person answering to describe in greater detail what is going on. He has to think about his answer. It often happens

that an operator who answers an open-ended question suddenly knows what the cause of the failure is, because he has to think everything through once more.
With open-ended questions, you don't put words in someone's mouth. You can use them to make sure that the answer has been formulated by the person themselves.

If you ask: "Does this happen often?" and someone answers "yes," you don't know how often something occurs. It's better to ask: "How often does this happen?"
Perhaps even more importantly, open-ended questions are usually not perceived as accusations. Compare the two questions below:

1. How did you set up the installation?
2. Have you configured the installation correctly?

The second question is much more readily perceived as an accusation or as a lack of trust.

Which Questions Can you Ask?

It's about determining exactly what is wrong with the equipment. You need to establish this as precisely as possible. A description like "It doesn't work?" or "That stupid thing is broken again" gives you very little to go on. To resolve the failure properly, you must first know exactly what the failure is.
If you consider a failure as a deviation from the normal behavior of an installation, you immediately have many starting points for asking more questions.

These questions may cover the following topics, among others:

- What is the normal behavior of the installation?
- What's different now, what's the failure?
- Why is this a problem?
- How serious is the failure?
- Where does the failure occur?
- Does the failure occur constantly or intermittently?

- When did the failure first occur?
- What has changed?

It is important not to jump to conclusions about what is going on, but to first map it out.
It is fairly easy to remember which questions you might ask if you memorize the list of question words.
In the following paragraphs you can read a number of questions you could ask. For your convenience, you will also find these summarized in Appendix C.

What?

You can ask many questions that begin with the word "what." The most open-ended question of all is: "What is your problem?" It is now important to make the description as specific as possible. We need to know exactly how the equipment deviates from normal behavior:
What is the asset doing wrong?
What do you notice about the output of a production machine?

A second important group of questions in the category 'what?' is the question of what exactly happened when the failure was first observed:

- What did they see happening?
- What did they hear?
- What did the operator do just before the failure?
- What did he do when he observed the first symptoms?

It is especially important to know whether the operator has already changed anything himself. Perhaps he has already replaced parts, cleaned, changed settings, or something similar. It is important to know what condition the asset was in when the failure occurred.

What is the Normal Behavior of the Installation?

In some cases where the repair technician knows a machine very well, this question is unnecessary. In other cases, it is very enlightening. Sometimes

a user sees or hears something that is completely normal, but considers it a failure.
Maybe the failure occurs in a detail of the installation that the maintenance technician is not as familiar with as the operator.
Resolving a failure is essentially returning a system to its normal behavior. Therefore, it is necessary to first know what normal behavior looks like. Often, the operator of a system, who spends the entire day with the system, knows this behavior in detail.

Why?

Why is this failure a problem?
This question is especially important for finding a solution quickly, even if it is only a temporary one. It is often not necessary to restore a piece of equipment to its original condition in order to solve the problem. Even if the root cause of the problem is not yet known, it is sometimes possible to address the effects.

Take the example of a leaking gearbox. If the quality control inspector is worried about a few drops of oil getting into the product, placing a drip pan can be a good solution. Although many people don't consider placing a drip pan a proper solution for a leak, it can be a good temporary fix and perhaps even a permanent one. With some gearboxes, a little seepage is nearly impossible to prevent. A drip pan can then provide good permanent protection. A complete overhaul is not always possible in the short term. The goal is not to restore equipment to its original condition. The goal is for the system as a whole to meet its requirements.

Another question with the word 'why' is: 'Why do you think that the asset isn't working properly?' So you ask the user for his or her idea about the cause of the failure. Many users know fairly well how their assets work and can help you find the cause.
You should never assume, however, that they are right. You always have to verify that. They can sometimes save you a lot of investigative work, though. Moreover, your collaboration with the operator will go much more smoothly if you listen to him and take his suggestions seriously. If

you as the expert immediately know that the cause the operator mentions is not the correct one, explain to him why that is the case.

How?

One of the questions you can ask is: "How significant is the failure?" This question has many variations. In case of deviations or damage to the product, you can ask how extensive the damage is. It is also important to know how many products show the deviation: all of them or only a few?
When there is a temperature deviation, you can ask how large the temperature deviation is. With a reactor that is slowly heating up, the question might be: "How much longer will it take now?" With the gearbox: "How large is the leak?"
Another question with the word 'how' is: 'How do you notice the failure?' It is possible that there is no failure in the actual process at all, but that a measuring instrument is deviating or that someone else reported the failure to your conversation partner, but that he has not yet observed the failure himself. If an operator, for example, reads a temperature that is too low on a screen, it may well be that the actual temperature is simply correct.

Where?

Where is the defect on the product? Where is the temperature deviation observed? Do the damaged products follow a different route through the production process than the undamaged products? Are they located, for example, in a different position in the machine? Where is the leak located? Where exactly do you hear that rattling sound? Where is the tear in the packaging?

When?

When does the failure occur? Occasionally or constantly? If a failure occurs only occasionally, does it coincide with other things? If it is a recurring failure: at what time or day of the week does the failure occur?
When did the failure first occur? What date? What time?

An important question is whether a failure occurred suddenly or deteriorated gradually. The latter points, for example, to contamination or wear of equipment or the calibration of instruments.

Which?

Which asset is experiencing the failure? In which rooms is the light not on, for example? Is that only in this room, or in the entire building, or perhaps in the entire city?
At which products do you notice the failure?
Which settings do you use?

Who?

When asking "who?", you must always be careful and communicate that you are not looking for someone to blame, but for a cause. If people think they are responsible for the problem, they will be less cooperative. Being blamed for the failure is not good for your reputation, for keeping your job, or for your status within the group. Always assume that operators are doing their best. Always be aware that approximately half of all industrial failures are caused by human error. Perhaps a manual is not clear, perhaps the operation of the asset is simply too complex, or users are insufficiently trained.
Nevertheless, it is sometimes necessary to ask the question "who?" For example: "Who observed all of this?" or "Who set up that machine? I would like to ask them a few more questions." In a number of cases, you discover this way that your conversation partner was not present at the start of the failure. Perhaps the previous shift reported the failure to him.

The Classic Car and Ice Cream Shop Story

A classic story for troubleshooters is the tale of a car that wouldn't start when a family bought vanilla ice cream, but would start fine with other flavors.
Once upon a time, there was a man who had bought a new car. His family had a tradition of eating ice cream as dessert every Saturday evening after dinner. But the ice cream shop carried several flavors, so after the meal the whole family would vote on which flavor they wanted that evening,

and then the man would drive to the store to pick it up. Ever since he had bought his new car, something strange had been going on. "Every time I buy vanilla ice cream, my car won't start when I come back out of the store. When I buy any other flavor, the car starts without any problem," he wrote in a letter to the manufacturer. "I know this sounds ridiculous," he continued, "but I'm completely serious: what is it about your cars that causes it not to start when I buy vanilla ice cream, yet starts without any trouble when I buy other flavors?"
The director of the auto plant was understandably skeptical about this letter, but sent an engineer to investigate nonetheless. The engineer was surprised when he was greeted by a successful, clearly well-educated man in a nice neighborhood.

They had agreed to meet each other after dinner. The two got in the car and drove to the ice cream shop. That evening they bought vanilla ice cream and indeed, when they returned to the car, it wouldn't start. The engineer came back three more evenings. The first evening the man bought chocolate ice cream. The car started. The second evening he bought strawberry ice cream. The car started. The third evening he ordered vanilla again. The car refused to start.

The engineer, being a logical thinker, refused to believe that this car was allergic to vanilla ice cream. He decided to continue his visits until he had solved the problem. He began taking notes: time of day, type of gasoline, driving time there and back, and so on. Soon he had a clue: the man spent less time buying vanilla ice cream than other flavors. Why? The answer lay in the layout of the store. Vanilla, being the most popular flavor, was kept in a separate freezer at the front of the store for quick service. All other flavors were at the back near a different counter, where it took considerably longer to pay.

Now the question for the engineer was: why did the car not start when it took less time to return? When time turned out to be the problem – not the vanilla ice cream – the engineer quickly came up with the answer: vapor lock. This problem occurred every evening, but the extra time needed to get other flavors gave the engine sufficient time to cool down, so that

the vapor bubble dissipated and the engine could start again. When the man bought vanilla, the engine was still too hot and the vapor lock could not dissipate.
And so the mystery was solved. Even the most seemingly absurd problems sometimes turn out to be real, and with thorough investigation and logical thinking, the solution can be found.

4.5. What, why, how, where, when, which, who is it NOT?

For all of the above questions, it is certainly just as important to ask where, when, or how a failure does NOT occur as where, when, or how it DOES occur. It may even be more important. To find the true cause of a problem, you must eliminate all potential causes that are not the actual cause. This can be done by looking for clear symptoms that occur in one location, time, or manner, but not in another.

The fact that an asset functions better at certain moments, or even functions properly, means that some components are working correctly. These do not need to be investigated further. The same applies to the fact that some parts of an asset are functioning properly. The light in the room you are in may not be working, but in the next room it is. This rules out many possible causes. It is therefore important to know what is NOT your problem. Precisely when there is a clear difference between what the failure is and what it is not, that is an indication of the possible cause.

4.6. Closed-Ended Questions

When you have mapped out the symptoms of the failure as thoroughly as possible with open-ended questions, you can start by asking closed-ended questions. You may already have an idea of a number of possible causes. With closed-ended questions, you can rule out some of them or at least clarify what happened.

> Important: It is always at least as important to know which symptoms a problem does NOT have as which symptoms it DOES have. This allows you to quickly rule out possible causes.

Closed-ended questions are questions such as:

- Is something leaking?
- Was it very hot in the production hall?
- Has that car been sitting idle for a very long time?

You can only ask these questions if you already have certain ideas about the cause of the problem. So, if the open-ended questions lead you to think that a controller is not functioning properly, you might ask whether there are any alarms on the controller's display.
With closed-ended questions you verify whether you have correctly understood the answers you received. But again: always ask the open-ended questions first, then the closed-ended questions.

4.7. Changes and Differences

One closed-ended question you should always ask is: "Has anything changed?"
Many failures don't just happen out of nowhere. Failures caused by a component reaching the end of its service life through normal wear are the exception rather than the rule. Many failures occur because something has 'changed'. Perhaps the speed of an asset has been increased, or the installation is being subjected to a 'heavier load' in some other way. It is also possible that slightly different raw materials are being used.
When many people think of the word "change," they think of clearly visible technical changes to an asset. Therefore, phrase your question in a few different ways. Also use words such as:

- improved
- optimized
- configured
- adjusted
- replaced
- maintained
- accelerated
- overhauled
- combined

- corrected
- repaired

Comparable to asking about changes is asking about differences. There may be a number of products, parts of products, or machines where the problem does not occur. So there is something different about those compared to the places where the failure does occur. Perhaps there are scratches on all the left front fenders in an auto plant. The question then is why there are scratches in that spot and not somewhere else on the car. It is therefore useful to pay attention to things that are or were different. The answers to the above questions will generally give you a good idea of what is going on and, in any case, a picture of where to look for the cause.

4.8. Length of the Conversation

The above questionnaire may seem a bit extensive, but if you imagine an ordinary conversation in which these questions are asked correctly, it need not take more than a few minutes. It is certainly worth covering all the topics, because asking questions always produces results much faster than searching and tinkering. In Appendix C you will find an overview of questions you can ask.

Always start by asking open-ended questions, then closed-ended questions for verification. Only then begin a detailed inspection of the asset itself, and try to resolve the failure only after you have found the cause.

Case Study: 7WH Analysis for Cooling Compressor C-15

Maintenance technician John has arrived and has asked Lisa a number of questions. Systematic questions to Lisa reveal important details:

What? Constant rattling noise from compressor unit that intensifies during startup, temperature gradually rising from normal 4°C to now 7°C
When? Both noise and temperature rise began Monday around 16:00, gradually worsening
Where? Sound coming from compressor unit outside, temperature problem in cold storage room inside

Changes? Monday coolant topped off by external service technician
Differences? Other cooling units (C-12, C-16) are operating normally

The pattern shows gradual deterioration, not sudden failure. The timing after maintenance and the specific rattling sound point to mechanical problems, possibly related to the recent service.

The questions have given you the information you need. Now you go on-site to verify this information and supplement it with your own observations. In Chapter 5 you will learn how to systematically carry out the initial reconnaissance.

MW01

5. The First Reconnaissance

In the previous chapter, you read about questions you can ask when someone reports a failure to you. It is often best to do this first, before going to the location of the failure. If you have a clear picture of the failure, you will know much better which tools, measuring instruments, or parts you need to bring. That can save you a lot of back-and-forth. But now it is time to go to the location of the failure.

5.1. Always Check in with the Responsible Persons

This can be the owner of an asset, the person who reported the failure, or for example the on-duty operators in a control room.
If they don't know you're there, very dangerous situations can arise, for example if they start up assets.
If you come forward and talk with them for a while, you also build a lot of goodwill. Even if it isn't strictly necessary for the repair of the failure.

After you have checked in, asked the right questions, and arranged the permits you may need, you can proceed to the location of the failure. As emphasized earlier, safety is crucial in failure work.
Often, the answers to your questions from the previous chapter alone are not yet enough to make clear what is going on. Then it becomes necessary to search for the cause of the failure. To find that cause, you need to know as many facts about the failure as possible. Even if you think you know what the cause of the failure is, you must verify whether you got it right.

From the condition in which the machine finds itself immediately after the failure, you can often infer a great deal about the cause. The more things are still in the same state as when the failure occurred, the more insight you can gain into what exactly happened.

It is therefore better not to change anything and first assess what you see when you arrive at the failure. This is not yet a detailed investigation, but consciously observe whether you notice anything unusual, such as leaks, alarms, or parts lying on the ground.

Make sure nobody changes anything. If you change certain things now, you will no longer know exactly what the installation looked like later. Once you have an idea about the possible cause, you can often verify whether this idea is correct by examining the installation more closely.

> Only act immediately if you can prevent something worse by doing so. There are naturally situations where you must act right away. With a major leak, you may need to close a valve. In other cases, allowing a machine to keep running could cause serious damage, and you must shut it down if that is the case. Always make sure you remember what you changed. Take notes and photographs.

5.2. Verify Information

The first thing you need to do is, if possible, verify that the information you have received so far is correct.

Wait a moment, you think, why do I need to verify the information: if I don't believe the answers, I might as well not ask the question? This is not entirely true. Your conversation partner has observed things, then interpreted them, told you about them, and you have tried to understand them. These are four steps that can lead to misunderstandings.

If you ask three people who were present at a particular event to describe that event, you will usually get three different stories. There are several reasons for this:

- People perceive things incorrectly. People count incorrectly. People overlook things, etc.
- Everyone interprets observations in their own way. Everyone thinks he clearly saw what happened, but always misses a few pieces of information. They unconsciously fill these in with what seems most logical to them. 'That's how it must have happened.'
- People don't like to admit that they don't know something. They sometimes give answers while they actually don't know at all.
- People want to keep things quiet. Sometimes they are ashamed or afraid of reprisals.

- Some descriptions are subjective: what I call a small scratch on a surface, for example, you might call a large scratch.
- People tell you things they haven't seen themselves, but that someone else told them.

The result of all this is that the story you have so far contains a reasonable amount of correct clues, but you still need to verify them.

A key category to verify are the answers that provide a clear distinction between where, when, or how the failure does and does not occur. Is it really true that all products from machine A are damaged and the products from machine B are not? There may also be a number of damaged specimens from machine B. Is it really true that the failure is occurring for the first time now?

A second important category to check are the subjective descriptions. These are descriptions from which you cannot determine how serious or how large something actually is. If someone tells you that a valve is hot, it could be lukewarm or so hot that it is glowing. If that person has not precisely measured the temperature, it is best to start checking yourself. There are many subjective descriptions. Check them as much as possible. How large is the scratch, really?
How does that scraping sound? Does it make a wavelike pattern or does it remain constant?

A third category to verify consists of statements that seem illogical to you. Some answers may not make sense to you or may not align with your experience. With others, you may have a feeling that something is off. Always trust your instincts in these cases and verify the statement yourself, if possible.

5.3. Systematic Observation

What you mainly do at this stage is look carefully, listen, and also smell whether anything unusual stands out. Are there any loose parts? Can you see that something is damaged? Do you smell the odor of scorched components in a control cabinet? Are there any indicator lights on? Are there any error messages? Do you see any leaks? Do you hear any strange noises?

So at this point you are not yet 'repairing' anything, changing parts or adjusting settings. You are only looking, listening, feeling, and smelling. We try to postpone the actual work on the machine until we are certain of the cause of the failure. Working almost always takes more time than observing things. It is also dangerous! You can damage the equipment you are working on or injure yourself. You also change a number of things on the installation by working on it, thereby erasing potentially important 'clues.'

5.4. Checking Basic Conditions

Many problems arise from human error. It is entirely possible that someone used the wrong settings for a process, or perhaps the wrong raw materials or components, or that a machine stops unexpectedly. Always check the basic conditions and settings of these assets to confirm that everything is as it should be.

Document the Original Situation

One of the most undervalued aspects of effective failure work is documenting the condition in which you find the equipment before you change anything. Too often, maintenance technicians go straight to adjusting settings or replacing parts, without documenting the starting situation. This regularly leads to situations where you no longer know how everything was originally configured, making it impossible to determine later whether your changes brought improvement or made things worse.

Always take photos of critical information before you begin: display screens with error messages, switch and button positions, hand valve positions, and visible damage or abnormalities. Your smartphone is well suited for this, as the photos are automatically time-stamped. Also record the values of important gauges such as temperature, pressure, current, or RPM. This baseline information is invaluable as a reference during your analysis.

Also think about documenting what you hear, smell, or feel. A short video recording can capture the sound of a defective machine, which can later be useful for comparing with the sound after repair.
The principle is simple: you cannot return to a situation you have not documented. Moreover, good documentation prevents after-the-fact disputes about "how it originally was" among the various parties involved. When an operator, production manager, and maintenance technician all have a different picture of the starting situation, effective problem-solving becomes nearly impossible.

Case Study: First Observations at C-15

At the compressor unit, Pedro systematically observes:

- Constant rattling noise, especially during startup
- Compressor runs more frequently than normal (every 8 minutes vs. the usual 15 minutes)
- Light oil stains under the unit, not previously noticed according to Lisa
- Temperature gauge cold storage cell reads 7°C, the orange warning light is on
- Condenser fan running normally, no blockages visible

The combination of more frequent cycling, oil loss, and rattling noise suggests internal mechanical problems. The unit is still functioning, but has to work much harder to maintain the temperature — a sign of reduced efficiency.

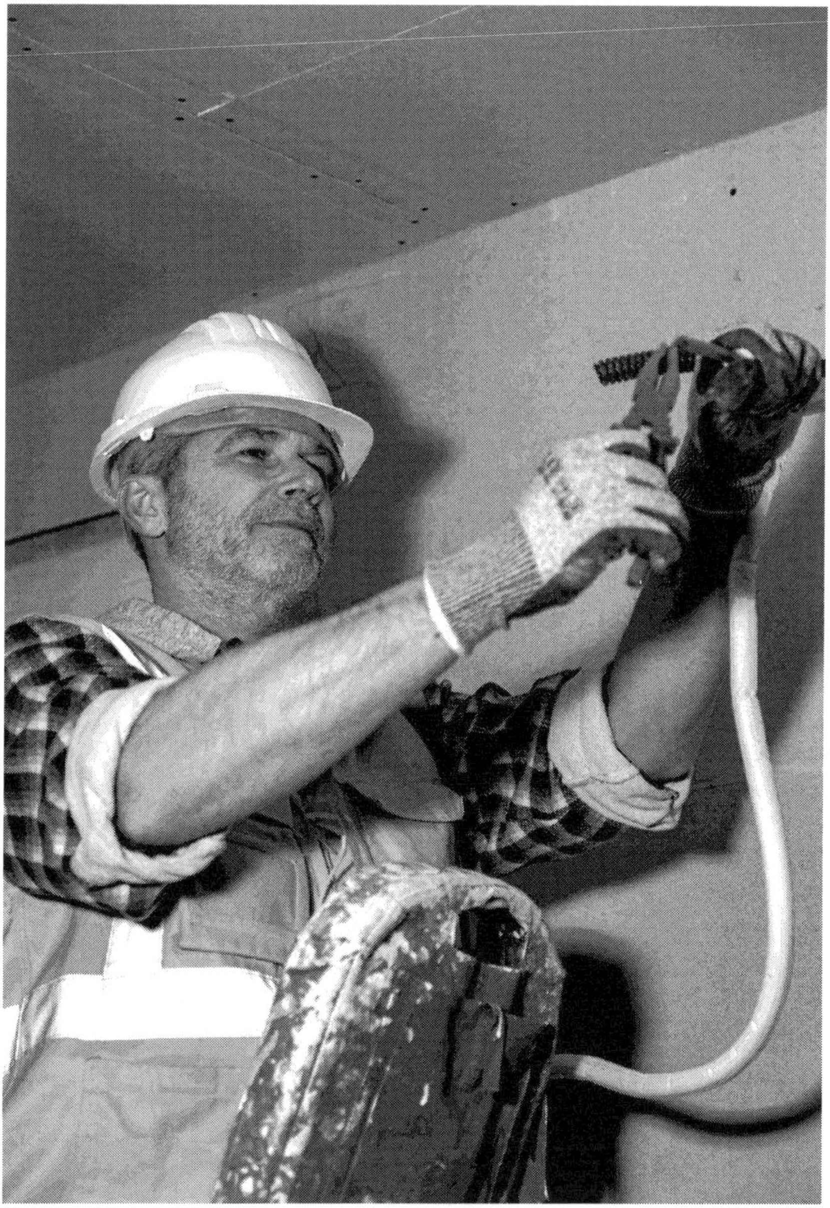

6. Identifying Possible Causes

Now that you have asked all the relevant questions about the failure to the operators of the equipment and have performed a visual inspection of the asset, you may already think you know what the cause of your problem is. If that is the case, because it is a failure that happens more often on this piece of equipment, you don't need to spend time on further analysis. If it happens regularly it may very well be the cause also this time. What you need to do first is think logically about whether the cause you have in mind explains all the symptoms that have occurred. So if a machine stops every time ten seconds after starting, your cause should not only explain that the machine stops, but also that it runs for ten seconds first. If that is the case, you can check the component or setting that is causing the failure and then repair it. After you have repaired it, you test it. If everything works correctly: congratulations, you have resolved the failure. However, if an asset has the same failure so often that you already know the cause without a thorough analysis, it is time to improve the asset permanently. This is the main topic of book 4 of this series: Root Cause Analysis in Maintenance.

In many cases you will know the cause and be able to resolve the failure, but what if you don't? In that case you need to make a list of all possible causes. There are several approaches for this. Which one works best for you depends on the situation and your personal preference. In this chapter you will read about all of them.
In any case, the first thing you need to do is figure out how the asset works. The better you understand how the asset functions and behaves under normal circumstances, the more likely you are to find the cause of the failure. If you know nothing about the asset, you first need to find out how it is supposed to work.

A good maintenance organization collects as much information as possible about its assets when they are running well. What processing steps does your product go through? What are the normal temperatures and pressures in your system? What sounds do you hear when the asset is in operation? What steps does the asset go through during a production cycle? How many amps does each electric motor draw? How do the controls work?

All this information must be stored centrally in the information system of the maintenance department. In any case, the failure technician must know the basic principles of how the asset operates, so that he can also understand the information in this system. A good failure technician will occasionally just watch a running machine for a while to see what that machine does during normal operation.

How well you know the asset that has failed determines your next step. There are several things you can do to find the cause of the failure:

1. Investigate what has happened to the asset in the last days or hours
2. Use the failure diagnostic screens of the asset
3. Read the user manual
4. Consult experts
5. Search the failure history
6. Look for differences
7. Perform measurements and inspections on the asset

Whichever approach you choose, you must ask yourself the following question:

"Is there a failure cause that explains why exactly this failure occurs in the described manner, and does the potential cause also explain why the failure does not occur at the places, times, etc. where we have established that it does not?"

The answer to the question must also explain why the failure does not occur at other locations, at other times, or with other products. You can see: the question of where the failure does not occur is just as important as the question of where the failure does occur.

If, for example, some of the products are damaged and another portion are not, then you need to ask yourself why exactly that portion of the products is damaged. What are the differences between the portion of the products that is damaged and the portion that is not damaged.

Often, a maintenance technician can list a number of possible causes based on his technical knowledge and experience. For example, he knows failures from the past involving these machines or other similar installations. The causes that come to mind first are often indeed the actual cause. However, it is necessary to ask yourself whether the cause that first comes to mind actually aligns with all the information you have.

6.1. What has been happening lately?

As you read earlier, assets that are operating stably rarely fail. Usually, a change takes place. This change can consist of many different things, such as:

- Heavier loading, for example due to the use of a different raw material
- A change in packaging materials or other materials used by them
- Performing maintenance or replacing parts
- Changing ambient temperature
- A failure of an asset that is connected to it

It is impossible to list all the changes that can occur, but I hope you understand what I mean by 'change'. If you look back a few days into the history of the asset, you will often find something that puts you on the right track toward the cause of the failure. Make sure you always know the recent history of the asst your working on.

6.2. Failure Diagnostic Screens

Many modern assets have a screen that displays failures, with the possible causes and what you can do about them. In some cases there is no screen, but for example only a few warning lights. In other cases you need to connect a laptop or similar device to the asset to read the failures. Make sure you have this equipment and that you know how to use it.

Digital manuals and online asset information can accelerate root cause analysis by providing direct access to technical specifications and known failure patterns. QR codes on assets can provide quick access to relevant documentation. These tools support your systematic analysis, but do not replace the fundamental logic of the methodology.

6.3. Read the User Manual

Many manuals have a section called 'what to do in case of failure.' This contains the most common failures, their causes, and how to resolve them. If you don't have the manual, do a quick internet search. Very often you'll find a manual within seconds. It's not always possible to find the exact right type of asset, but if you find one from the same series, from the same manufacturer, there's a good chance that what it says also applies to the asset you're working on.

6.4. Consult Experts

Now that you have all the information about the failure at a glance, a phone call with the supplier or manufacturer of the system can in many cases yield very valuable information. The employees of the service department of a specialized supplier have a great deal of experience with exactly that type of installation and can already provide many useful pointers over the phone.

Don't hesitate to ask colleagues for help. There may be someone who knows the asset better or who has had to solve the same problem before. Just make sure that person explains to you how it works and how you can fix it. If he goes ahead and solves it himself, you won't learn anything from it and you'll have to ask him for help again next time.

6.5. Search the Failure History

If there are records of past failures, read them. A large portion of failures in technical installations occur more than once, even if only every few years. If the previous time the exact cause of a particular failure was already properly determined, then that is now also one of the most likely causes.

To be able to do this, it is necessary that everyone writes good failure reports.

6.6. Look for Differences

If there are two assets that are identical and should be doing the same thing, and one works fine while the other does not, you can systematically compare one with the other to determine what their differences are. Those differences will exist. Always. Engineering doesn't work by magic. If one asset works and the other doesn't, there is a difference.
Of course, this difference doesn't have to be in the asset itself. It can also lie in the settings, raw materials, the ambient temperature, or other factors.

Case Study: 5M Analysis for Cooling Compressor C-15

Based on what he has observed so far, John compiles a list of possible causes:
Man: Service technician topped off coolant on Monday – possible assembly errors?
Material: Wrong coolant type or contaminated fluid added?
Method: Procedure for topping up possibly not followed correctly
Medium: Warm day (28°C outside), but other units have no problems
Machine: Internal compressor damage, system leakage, or defective valve

Three main hypotheses remain after critically testing whether these causes explain all the symptoms: a leak causing the system to run under negative pressure, incorrect coolant that is foaming or causing corrosion, or mechanical damage to the compressor during service. All three explain why the problem started after maintenance and is gradually getting worse.

> **Letter-F Exercise**
> Count how many times the letter "F" appears in the text below:
> **EFFECTIVE METHODS OF**
> **PROFESSIONAL TEAMS ARE THE RESULT**
> **OF CAREFUL PLANNING AND**
> **THE BENEFITS OF EXPERIENCE**
> I'll explain why in a moment.

6.7. Analyze the Operation of the Asset

If you know the operation, normal behavior, and history of the asset well, you can analyze the cause of the failure yourself. The central question in the analysis is:

'Does the potential cause precisely explain that a failure occurs in the described manner, and does the potential cause also explain why the failure does not occur at the places, times, etc. where we have established this?'

You have read this question before, but it is essential for solving a failure quickly and correctly the first time. There are even people who say that if you have properly inventoried all the symptoms of your problem, there is only one possible cause that precisely explains why that failure occurs with that product, at those times, in that location, and in that manner. In my experience, there can also be more possible causes, but you can always cross off a large number of possible causes that you no longer need to investigate, and a large number of components that you do not need to check, simply because they do not explain the symptoms of the failure.

Do this first at a high level, focusing on the main sub-units of the installation, and then in more detail at the component level. By 'sub-units' I mean units such as the control system, the drive, the heating system, and the fuel supply. You may not yet know whether the problem lies in the fuel pump or the fuel filter, but you may be able to tell from the symptoms that your asset is not receiving fuel. So always try to rule out as large a part of the installation as possible as the cause of the problem. Ultimately, one or a few possible causes will remain. At this stage it is very important to reason logically and to be consistent. Once you have established that the cause cannot lie in a particular part of the installation, do not waste time investigating or measuring that part.

What you then need to do is think about how the asset should function. Every asset performs its 'tasks' in a number of steps. Map these out systematically:

- With an electrical circuit, you can follow the path that the electrical current travels through your wires and components. Use the electrical schematic.
- In the case of a liquid flow or powder flow, you could follow this flow through your pipes and equipment. Use the Piping and Instrumentation Diagram (P&ID).
- You follow the product through the machines along a production line.
- With software, you could trace the processing of information throughout the entire program.

Try to understand precisely at each step what is happening and how you can measure whether your product, fluid, or electrical current has the correct properties. For example, does a certain part of your circuit need to have voltage? How much? Positive voltage or negative?

It often helps your understanding to draw out your own process schematically. This doesn't need to happen 'formally' at all. The more relaxed you are, the better you think. When you're too tense, you tend to get stuck in the same thinking pattern you're already in. You might only think of two possible causes and no longer come up with new ideas. When you relax, new ideas come to the surface easily. I've heard many times from maintenance technicians, and experienced it myself, that you can spend hours struggling with a problem and simply can't find the solution. When you let the problem go and you're sitting in your car on the way home from work, the solution suddenly comes to you.

That is why it often works even better to scribble something on a napkin than to produce a formal drawing of your asset. The point is that you truly understand how the asset works or should work. When you draw a diagram or sketch of the asset's operation yourself, you force yourself to structure your thinking and are less likely to overlook details than when you reason without a sketch or diagram. You may discover that it would make sense to place a valve or another component somewhere. If you were to study an existing drawing, you might overlook those elements, even though they could be essential for explaining the cause of the failure.

In the text in the box, in which you just counted how many times the letter "F" appeared, the correct answer was eight. Most people only count five or six. This is an example to illustrate that you can easily overlook something when reading a drawing or manual while already thinking you know what it should say.

Once you understand how your asset should work, you can begin measuring.
Measuring only makes sense if you know which measurement result is good and which measurement result is bad. To find the cause of your failure as quickly as possible, you should try to rule out as many parts of your system as possible as the cause of the failure. You generally do this by checking the inputs of the main groups of components that make up your system.
If you are working on a complex system with many components or steps, it is best to start in the middle. Suppose you are measuring a circuit with 256 components. If you start in the middle, at number 128, you can already rule out 128 components as a possible cause after your first measurement. Next, you measure at the middle of the section with the broken component, so perhaps at component 64 or 192. If you can divide the remaining group exactly in half each time, you can find the defective component in eight measurements. If you started with component 1, then 2, and so on, you would need 256 measurements in the worst case.

It is of course not always possible to start exactly in the middle. But if, for example, an electric motor is not running, you can measure whether the motor is receiving electrical current. If it is, something in the machine may be stuck or broken. If it is not, there is something wrong with the control system or power supply of the motor.

Often a system consists of a few main groups, such as a control cabinet, wiring, an electric motor, and a pump. It is then most logical to check at the boundary between two main groups whether everything has been working correctly up to that point. Is the pump being controlled properly?

A variant of 'measuring' is applying a good signal somewhere halfway through the chain of components. If the output signal of the entire system is good, the problem most likely lies in the first half of the chain. If the output signal is not good, the problem lies in the second half.

Sometimes a problem or its symptoms have already passed by the time you start measuring. A system can fail due to a brief signal that disappears immediately. In those cases, it can be useful to record the relevant data so that you "catch" the problem when it occurs again. This can be done with a data logger, monitor, or, for example, a camera.

6.8. Safety During Measurements

Pay close attention to your own safety and that of others when taking measurements. Working on assets can be extremely dangerous. Contact with electricity, steam, or chemicals can cause serious injuries. Moving machinery, loose parts, suddenly moving parts, and falling parts can also be extremely dangerous.
Unfortunately, people still die every year while performing maintenance. Working on failures is especially dangerous, because you cannot prepare for them days in advance, and there is always a sense of urgency to get things running again.

75% of industrial accidents
occur during maintenance activities!

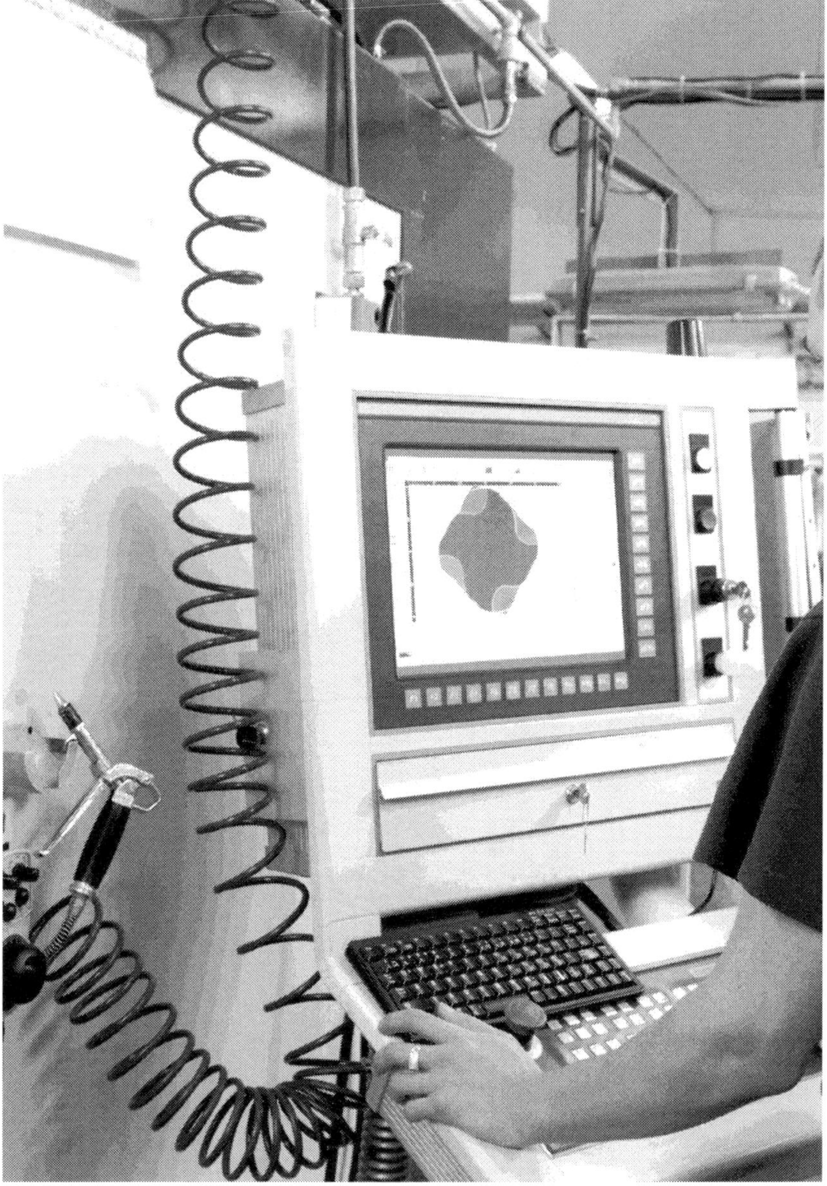

7. Digital Tools

7.1. Smartphone as a Diagnostic Tool

Your smartphone is much more than a communication tool during breakdown work. Modern smartphones contain sensors and computing power that until recently could only be found in expensive specialized measuring equipment. By using the right apps, you can turn your phone into a versatile diagnostic tool that you always have with you.

The camera on your smartphone can, for example, perform vibration analysis by recording slow-motion video. If a machine vibrates irregularly, you can capture this and later analyze it frame by frame to discover patterns. The microphone can perform sound analysis in which different frequencies are isolated, which helps in identifying bearing sounds versus gear sounds. Many apps can also measure sound levels and track trends.

For temperature measurement, you can connect your smartphone to inexpensive infrared accessories that create thermal images. This helps in quickly identifying overheated components, poor electrical contacts, or blocked cooling channels. The GPS function can be useful for recording exactly which assets and locations you have inspected, especially in large plants where assets are not always correctly labeled.

7.2. Digital Measuring Instruments and Real-Time Monitoring

Modern assets are often equipped with sensors that continuously collect data on pressure, temperature, vibrations, power consumption, and other parameters. This information is usually accessible via SCADA systems, web interfaces, or vendor-specific apps. Instead of taking measurements manually, you can use this historical data to identify patterns not detectable by the human eye.

Look at the trend in a motor's power consumption in the days leading up to the failure, for example. A gradual increase can indicate rising friction due to wear or contamination. Sudden spikes can indicate electrical problems. Temperature patterns can show whether a cooling system is gradually becoming less effective, while pressure measurements can reveal hydraulic or pneumatic problems.

For more complex analysis, you can export this data to your laptop or tablet and use spreadsheet software to create graphs that show trends more clearly. This helps especially with intermittent failures where the problem is not present at the moment you arrive.

Figure 7-1 An infrared camera makes temperatures visible

7.3. Augmented Reality and Remote Support

One of the most promising developments is the use of augmented reality for remote support. With a tablet or smartphone, you can video-call a specialist who can watch in real time what you are seeing. The specialist can place digital annotations on your screen to point out exactly which component he means or which measurement you need to take.

This is especially valuable for specialized equipment where local expertise is lacking. Instead of flying in an expensive specialist, that specialist can provide remote support while you perform the physical actions on-site. Some suppliers offer this as a standard service.

For safety, you always remain personally responsible for performing the LMRA and taking safety measures. The remote specialist can advise on technical aspects, but you know the local safety situation best.

7.4. Predictive Analytics and AI Support

More advanced systems use artificial intelligence to identify patterns in sensor data that indicate impending failures. These systems can warn you that a certain component will likely fail within a short time, based on subtle changes in vibration spectrum, temperature profile, or other parameters.

As a breakdown technician, you don't need to understand or configure these systems, but it helps to know they exist. When such a system issues a warning, you can take preventive action before an actual failure occurs. This saves not only time, but also prevents the stress and risks of acute failure situations.
When analyzing a failure, it can also help to ask whether the asset has recently generated AI alerts. There may have been subtle changes that went unnoticed by humans, but that are nonetheless relevant to your 5M's analysis.

7.5. Digital Documentation and Knowledge Management

QR codes on assets can provide direct access to digital documentation, failure history, maintenance reports, and technical specifications. Instead of walking to an office to look up documentation, you simply scan the code and have immediate access to all relevant information on your smartphone or tablet.

Some companies use digital logbooks where every action performed on an asset is automatically recorded. You can then see exactly who performed maintenance and when, which parts were replaced, and which settings were changed. This helps enormously in identifying recent changes that may have caused a failure.
Also make use of digital documentation yourself by taking photos of settings before you change them, or by recording short videos of abnormal

behavior. This documentation can later help when writing your failure report and can be valuable for colleagues who encounter the same problem.

7.6. Integration with the 5-Step Methodology

All these modern tools support the 5-step methodology, but do not replace it.

In step 1 (Describe), smartphone sensors and digital logging help accurately record symptoms. Video recordings can document behavior that is difficult to describe in words.

At step 2 (Possible causes), historical data and AI warnings can generate additional hypotheses alongside your 5M's analysis. Remote experts can suggest alternative causes based on their experience with similar systems.

During step 3 (Eliminate), digital measuring instruments make it possible to quickly and accurately take measurements that can confirm or rule out causes. Real-time monitoring can show whether an intervention has an immediate effect.

For step 4 (Repair), remote support can help with complex procedures, while augmented reality can overlay step-by-step instructions onto your work area, and for step 5 (Follow-up), digital reporting tools make it easier to create comprehensive documentation with photos, measurements, and trends.

7.7. Practical Implementation

Begin simply by installing a few useful apps on your smartphone for sound analysis, vibration measurement, and temperature monitoring. Learn to use these tools during routine inspections, so you are familiar with them when you need them during an acute failure.

Ask your employer about the availability of tablets or laptops for breakdown work and inquire whether suppliers offer remote support. Many companies have already made investments in digital infrastructure without maintenance technicians being aware of this.

Remember that technology is a tool, not a replacement for systematic thinking. The fundamental logic of the 5M's methodology remains essential, but modern tools can help you apply this methodology faster and more accurately. Experienced technicians combine technical intuition with digital support for optimal results.

8. Eliminating Root Causes

In chapter 6 you learned about a number of methods you can use to create a list of all possible causes of your problem. All of these causes, except one, will not be the actual cause.
Eliminating causes is the most critical phase of failure resolution. This is where the difference between systematic work and costly trial-and-error methods is determined. Many maintenance technicians make the mistake at this point of moving too quickly to repair, which results in hours of wasted time replacing good parts.

8.1. The Core Question of Effective Failure Diagnosis

The fastest and safest way to eliminate causes from your list is usually by asking yourself the following question:

Does this cause explain ALL the symptoms that occur?

This may be the most important question in troubleshooting failures. In my thirty years of experience in the field of Maintenance & Reliability, I have seen time and again that failure technicians who apply this question rigorously are on average 50% faster than colleagues who skip this step. They also replace significantly fewer unnecessary parts.

Why This Question Is So Powerful

If you carefully consider all the symptoms you have listed — such as when the problem first started, when it occurs, what kind of alarms appear, where it does and does not occur — you can usually conclude that only one, two, or three causes can be the actual cause of your failure.
The word 'all' in the question is crucial. A potential cause must not only explain that, for example, a machine damages the product, but also that it happens at certain times and not at others, or that it only causes damage in one place. If a potential cause does not explain all the symptoms, it cannot be the true cause of your problem.

A Concrete Example

Imagine: a packaging machine consistently stops after exactly ten seconds of running.

Possible cause: "The main motor is defective"

Check this against the core question: Does a defective motor explain all the symptoms?

- Does it explain why the machine stops? Yes
- Does it explain why the machine runs normally for ten seconds first? No
- Does it explain the exact ten-second timing? No

Conclusion: A defective motor cannot be the root cause.

Actual cause: A timer setting that is incorrectly configured

- Explains why the machine stops? Yes
- Explains why after exactly ten seconds? Yes
- Explains why the machine does start? Yes
- Does it explain the repeating pattern? Yes

This cause explains all observed symptoms and is therefore highly likely to be the correct one.

8.2. Common Thinking Errors

Failure technicians regularly make these mistakes when evaluating causes:

"This cause explains the most important symptom" - Wrong. All symptoms are important, not just the most obvious one.

"This seems like the most logical cause" — Wrong. Logic without verification against all facts leads to assumptions.

"We're trying this first, because it's easy to replace" - Wrong. Ease of repair is not a criterion for the likelihood of the cause.

The correct approach is: "This cause explains every observed symptom, every time pattern, and every location where the problem does or does not occur."

When to Apply This Question

You don't have to wait until you've completed the list of possible causes to ask yourself this question. You can ask it immediately as soon as a possible cause is added to your list by the methods we discussed in the previous chapter. Many causes can be eliminated through logical reasoning, without taking measurements or touching components.

This method of elimination by reasoning takes less time than performing measurements and is also safer for you and your equipment, because you no longer need to work on the installation more than is necessary.

Once you have reduced the list of possible causes through reasoning to only a few that explain all the symptoms, it may be possible to verify which cause is the single true cause.

8.3. Verification of Causes

In an electrical circuit it is often possible to measure current or voltage to verify whether it is what it should be.

In software, it is often possible to devise a test to see whether a specific input leads to the expected output.
In mechanical systems or chemical processes, there are a few methods that are commonly used. The first is measuring parameters such as temperature or pressure at specific steps in the process. Another way is to perform a test to see if the process responds as expected.

In all cases, you must reason: "If cause A is the cause and we perform action B, the result must be C." This could be something like: "If the scratches on my product are caused by something protruding from the side of my machine (A) and we insert the product backwards (B), the scratches should appear on the other side of the product (C)."

8.4. Choice with Multiple Causes

It can happen that after all the reasoning and checks, more than one cause still remains.
In that case, you need to decide which possible cause you will address first. In some cases, it may be better to resolve all of them at once. For example, you may need to stop, start, and open a machine to address each possible cause, while the actual resolution of each cause takes only a few seconds. In that case, you might choose to resolve all of them simultaneously. For instance, if you know that one of the spark plugs in a twelve-cylinder engine is defective, it is often better to replace all twelve at once than to restart the engine twelve times.

In most cases, however, it is best to first try to resolve one cause, test it, and then move on to the next. This can be faster, but it will also teach you what the actual cause was, so you can think about actions to prevent the same problem in the future.
Be careful with swapping parts by trial-and-error. You should always perform the analysis first to limit the number of suspect components as much as possible. Replacing parts that are not actually broken can be a very expensive method due to the high cost of some components. You also run the risk that a defective component is caused by something else and that the good part you install will also fail immediately. This can be very confusing and can quickly deplete your spare parts inventory as well.

A variation on replacing parts one by one is removing parts one by one. Control systems and communication systems often consist of various components that operate independently of each other but influence one another. In those cases, there may be one component or input that disrupts the operation of the system as a whole. If you disconnect that component or input from the system, the system may start functioning better again. A communication link can contain too many boxes, incorrect cables, too many repeaters, or cables that are too long. This method can also be used to identify improper grounding.

Prioritization of Causes

If you have a shortlist of several causes that could be causing your failure, it is usually wisest to address one cause at a time. How do you choose where to start? There are a number of reasons to address a particular possible cause first.

The first reason to start with a particular cause is that it is the simplest and fastest to resolve. If you only need to replace a small part or reset something to eliminate one cause, and it would take you four hours to eliminate another cause, your choice is simple.

The second reason to choose a particular cause is when it involves small components that move a lot, especially when they are installed in a harsh environment: cold, hot, humid, dusty, vibrating, etc. These small components fail much more often than the "larger" components deeper within the asset.

If a few possible causes remain that both require a lot of effort to resolve, consider how likely the cause is. One way to do this is to reason: "Will this cause always lead to this problem, or do a number of other conditions need to be met to produce these symptoms?" The more conditions that must be present to cause the symptoms, the less likely the cause is. In other words: some causes lead to a failure on their own. Others only under special circumstances — for example, if additional failures occur as well. The more of these "special circumstances" that must be present to cause a failure, the less likely the cause is.

8.5. Recognizing and Breaking Tunnel Vision

One of the biggest pitfalls in troubleshooting failures is tunnel vision: clinging to one favorite theory while becoming blind to other possibilities. This happens especially under time pressure, when you want to find a solution quickly and latch onto the first plausible explanation that comes to mind.

Tunnel vision manifests itself in various ways. You notice that you keep repeating the same measurement, hoping that this time it will give the result you expect. Or you only look for evidence that supports your favorite theory, while ignoring contradictory signals. A classic example is the "experience trap": because last week the same problem was caused by a defective sensor, you automatically assume it's the sensor again, without analyzing the current situation.

Another dangerous form of tunnel vision is confirmation bias: you interpret all observations in a way that confirms your original theory. If you think the problem is electrical, you start explaining mechanical noises as "electrical contacts sparking" instead of objectively looking at other possible causes.

There are techniques to break through tunnel vision. The simplest is to take a deliberate five-minute break when you notice you're getting stuck. Step away from the machine, grab a cup of coffee, and come back with a fresh perspective. Often you'll then notice details you overlooked before. Another effective method is consulting a colleague — not to hand off the problem, but to get a second opinion. Briefly explain what you have found and ask: "What do you see that I might be missing?" A fresh set of eyes can identify patterns that you may have stopped noticing due to your focus on a single theory.
Try to deliberately consider a different M from the 5M's methodology. Often the solution lies in an area where you haven't yet searched systematically.

A powerful technique is to play "devil's advocate" against your own theory. Ask yourself explicitly: "What if my hypothesis is completely wrong? What other explanations are possible?" This forces you to step outside your existing thought patterns.
Above all, avoid changing multiple things at the same time. Under pressure, there is a tendency to "quickly try this as well," while your first intervention has not yet been tested. This leads to complete chaos in which you no longer know which change produced which effect.

Don't be too quick to settle for a "seems to work" result either. Tunnel vision can also cause you to see the first small improvement as proof that your theory is correct, while the problem has only been partially resolved. Always test thoroughly before concluding that the failure has been resolved.

8.6. Warning About New Components

One thing to be aware of is that new components often have a higher probability of failure than old ones. Yes, you read that right: new components often have a higher probability of failure. Many components fail due to improper installation or manufacturing defects. Therefore, never think: this part is new, so it cannot fail.

Case Study: Eliminating Root Causes at C-15

Test of core question: "Does this cause explain all the symptoms?"
Wrong coolant: Would cause immediate problems, not gradual deterioration. Eliminated.
Mechanical compressor damage: Would produce a constant loud noise. Our noise is soft and only amplifies during startup. Eliminated.
Leakage in cooling system: Explains all symptoms perfectly:

- Oil stains indicate a leak in the compressor oil circuit
- Gradual pressure/fluid loss explains reduced cooling capacity
- Compressor must cycle more frequently to maintain temperature
- Rattling noise due to insufficient lubrication caused by oil loss
- Problem after service: possible line damaged during work

Pressure test confirms: system loses 0.3 bar per hour. Leak found at fitting that was touched during service.

9. Further Causes and Consequences

Once you have found the cause of your problem, think about how this could have happened. What is the cause behind the cause?
If, for example, a bearing seized up and was replaced, ask yourself why it failed. If the bearing is equipped with a lubrication system, was it functioning properly? Was the bearing properly lubricated? If the bearing was not properly lubricated, was there perhaps something wrong with the lubrication system? You are not yet finished with your work. You will first need to ensure that the lubrication system is working properly again. If you don't, the new bearing will fail again quickly.
If a fuse had blown, why? It goes without saying that you need to look for the underlying causes of the problem to prevent it from recurring. The underlying causes you find also have causes. You need to find those as well. You need to keep searching until you arrive at a cause that enables you to solve the problem permanently.

9.1. The 5-Why Rule

Some people think that all problems have a root cause, an ultimate cause where the chain of cause and effect ends. In most cases, that is not true. Take the example of the bearing. The chain of cause and effect could then be:
Why Did the Machine Fail? — The Bearing Seized
Why? (2) - Insufficient Lubrication
Why? (3) - There was no grease in the reservoir of the automatic lubrication system
Why? (4) - It Was Not Refilled on Time
Why? (5) – The operator didn't know he had to replenish the grease
Why? (6) - The person who normally performs this task was sick and his replacement did not know he had to replenish the grease
Why? (7) – There is no formal procedure for it
Why? (8) – Work Pressure

It should be clear that we could go on like this for quite a while.

It is important to think far enough to reach the cause that can best be influenced. In this case, that will mean filling the grease reservoir and pointing out to the operators and their supervisors the necessity of keeping this reservoir filled. A good rule of thumb is that for each cause you search five levels deep for the root cause. So for each cause you must ask: "why does this cause occur?" and do this five times.

9.2. Vertical thinking

This search for underlying causes is further referred to as vertical thinking. Another form of vertical thinking is reflecting on the consequences of the failure. You may still need to take measures before startup to prevent further damage. These can be simple things such as cleaning up aggressive substances that can cause damage or cleaning up leaked oil that makes the floor slippery. In the case of the empty grease reservoir, you may need to lubricate a number of other bearings before starting up.

Horizontal thinking means asking yourself whether the same problem could also occur in other comparable assets in your organization. You can read more about this in chapter 13.

Case Study: 5-Why Analysis for C-15

Why Is the System Leaking? Fitting Came Loose After Service Work
Why is the fitting loose? Not correctly tightened after refilling coolant
Why not tightened correctly? Service technician had the wrong tool (wrench too small)
Why wrong tool? New fitting type since last year, old tool set not updated
Why wasn't the tool updated? Communication between the maintenance department and the external service provider is lacking

Horizontal thinking:
Check the other cooling units (C-12, C-16) for the same fitting type. Preventively inspect all fittings that were touched last week.
Follow-up Actions:
Repair the leak, refill and bleed the system, establish a communication protocol between the internal maintenance department and external service partners regarding equipment updates.

You have now identified the underlying causes. The next step is to safely and effectively repair the failure. In Chapter 9, you will learn how to prepare and carry out the repair.

10. The Repair

You will not read in this book how a specific failure should be repaired technically. That is a matter of craftsmanship in a particular field. It may require mechanical skills, electrical skills, or perhaps you need to be a software expert. It is also different for every asset. Nevertheless, there are a number of things you need to take into account before you begin a repair.

10.1. Definitive or Temporary Solution?

Of course, every failure must ultimately be permanently repaired, but in a number of cases it is better to first opt for a temporary solution. For example, you can equip a valve that cannot be operated automatically with a lever to enable manual operation, or you can repair a leaking water pipe by clamping a piece of rubber against it.

Sometimes you don't have the right parts available to resolve the failure definitively. In other cases, the final repair takes too long. In those cases, it is sometimes better to opt for a temporary solution.
When choosing a temporary solution, it is of course very important to carefully think through the consequences of that temporary solution. The safety or quality of the product must not be compromised. It is best to discuss with your customer, such as a production manager, how long you expect each option will take and what the further consequences will be.

Most customers or production operators appreciate receiving an estimate of this. They can also best assess whether the time is available for a permanent repair. They can therefore

Make sure the failure is permanently repaired as soon as a suitable opportunity arises. Therefore, always clearly indicate on your failure report that you have performed a temporary repair and that the permanent repair still needs to be done, or create a work order for it, depending on the procedure within your company.

10.2. Removing the cause or removing the effect?

Sometimes it is not possible to eliminate the cause of a problem. In that case, it can also be sufficient to eliminate the consequences. In the previously mentioned example of a leaking gearbox, placing a drip tray to catch the oil is such a solution. The solution that eliminates the consequences can be temporary or permanent. The company may decide that a little seeping from the gearbox is unavoidable. In that case, one may choose to install a permanent drip tray under the gearbox.

10.3. Preparing the Repair

Performing a repair is rarely routine work. That is why it is necessary to prepare the repair properly. As a maintenance technician, you need to think about how you want to carry out the repair.

- Which parts do you need?
- What tools do you need?
- What safety measures do you need to take?
- What problems might you encounter? What can go wrong?
- Et cetera

Parts and Tools

It obviously makes little sense to disassemble something and then discover that you don't have the parts or tools to put it back together. Therefore, think ahead as much as possible about what you will need. Bring it with you to the asset. This can save a lot of back-and-forth trips to a warehouse. Sometimes you don’t know exactly which parts are in a machine or which tools you will need. In that case it is better to have a little too much with you than to be just one part short.

Always make sure to reinstall the same parts or parts with exactly the same specification.
Remote expert support via video call can be valuable during complex failures, where a specialist thinks along remotely without being physically present. This technology can be especially useful for specialized equipment where local expertise is lacking, but always requires local execution of safety measures.

Safety

75% of accidents in industrial facilities occur during maintenance activities, as you have already seen. Of all maintenance activities, troubleshooting failures is the most dangerous. More dangerous than, for example, performing preventive maintenance. Failures always occur unexpectedly, so the repairs are never planned in advance. You may need to:

- move or lift heavy components
- remove guards;
- open pipes;
- climb on top of or under assets;
- remove the covers of electrical cabinets;
- remove the insulation from very hot components;
- and so on.

You are therefore performing hazardous operations. You must always comply with the safety rules of your company, the law, and the manufacturer of the asset being repaired.
Often, all kinds of people apply pressure to resolve the failure quickly. Don't let yourself be rushed. The most dangerous words when working on assets are "hurry up!" and "quick fix."

Legal regulations and company procedures exist for a reason: always follow them. But blindly following these procedures alone is not enough. In each individual case there may be specific risks that general regulations do not cover. So take the time to think things through first, and never do anything without knowing it is safe.

Last Minute Risk Analysis (LMRA)

Always perform a Last Minute Risk Analysis (LMRA) yourself.
This LMRA is best performed together with the operator. In any case, it's a good idea to do it with two people, to prevent it from being done too quickly and carelessly. During that LMRA you carry out three steps:

1. Identify hazards

Ask yourself which hazards can occur, despite measures you may have already taken. Ask yourself questions such as:

- Can I come into contact with hazardous substances?
- Can I come into contact with live components?
- Can I come into contact with hot components?
- Can I trip?
- Can I fall from somewhere, such as a platform?
- How heavy are the parts I need to move? Can I lift them?
- Do I have the right tools?
- What do I do if something goes wrong? Is there anyone nearby to help?
- Has all energy been removed from the asset? Think of electrical energy, springs that are tensioned or compressed, static electricity, hot components, pressurized lines, et cetera.

2. Determine Measures

Ask yourself what measures you need to take to eliminate the hazards from the previous question. You may need personal protective equipment, or you may need to cordon off a work area so that others are not put at risk. Always use LOTOTO before you begin working. LOTOTO stands for Lock Out, Tag Out, Try Out. Many companies have established procedures for performing LOTOTO on specific assets.

First, map out how you can secure the components you will be working on. Many companies have analyzed this in advance for their most hazardous equipment. They have documented how to ensure that, for example, hazardous substances, electrical currents, or pressurized gases cannot be released. Study that standard analysis, make sure you understand it, and verify that no changes have been made to the asset after the standard LOTOTO analysis was established.

LMRA CHECKLIST For every job – preferably with two people		
STEP 1: IDENTIFY HAZARDS **Energy sources:**	**Physical hazards:**	**Work environment:**
STEP 2: DETERMINE MEASURES **Plan LOTOTO:**		
☐ **Lock Out – Which energy sources to isolate?**	☐ **Tag Out – Where to attach warning labels?**	☐ **Try Out – How to verify system is de-energized?**
	Additional measures required: ☐ Personal protective equipment? ☐ Cordon off work area? ☐ Lifting aids? ☐ Flush / disconnect pipelines? ☐ Remove fuses from main circuit? Other measures?	
☐ **STEP 3: IMPLEMENT MEASURES**		
✓ START ONLY WHEN ALL STEPS ARE COMPLETE **If in doubt: STOP and seek advice** *Safety always takes priority over speed*		

Figure 10-1 LMRA checklist

Often it is not sufficient to simply switch a maintenance switch or turn a valve. In many cases it is necessary to flush pipelines and disconnect them completely. An electrical maintenance switch is often located in the control circuit of a circuit. In that case it may be necessary to pull the fuses from the main circuit. Never do this if you have not been trained and authorized to do so by the management of your company.

3. Implement safety measures

Carry out the safety measures before you begin your work:

- make the asset as voltage-free as possible;
- use lifting aids;
- make sure there is someone nearby;
- tie a ribbon to cordon off your work area;
- and so on.

Only when you are absolutely certain that you can work on the equipment without danger to persons, products, or equipment may you begin the repair.

10.4. Management and Safety

I know several companies that have succeeded in significantly reducing the number of accidents by consistently performing an LMRA for every job. They require their employees to complete this LMRA in writing, on a standard form. It is management's responsibility to regularly verify that this is actually being done.
Management should never ask after a failure has occurred: "Why did it take so long to repair?" That puts pressure on people and creates an unsafe work environment. The question a manager should ask is: "How did you make sure you could work safely? Show me your LMRA."

Typical Hazardous Situations in Breakdown Work

Failure work carries specific risks that differ from planned maintenance. By recognizing these risks, you can better prevent them. As a maintenance technician, you regularly find yourself in situations where the combination of time pressure, unfamiliar systems, and acute problems can lead to dangerous moments.

Work Pressure and Haste — The Biggest Risk Factor

"Hurry up, production is down!" These are words you often hear as a maintenance technician, and they represent the greatest threat to your safety. Under time pressure, you skip safety steps, forget tools and parts — forcing you to make extra trips back and forth — and you take

unnecessary risks with the mindset of "just this once without a climbing harness." On top of that, under pressure you make cognitive errors that lead to incorrect diagnoses, ultimately costing you even more time.
The most important thing you can do is explain that five minutes of extra preparation ultimately saves time. A systematic approach is always faster than an accident, and the 5-step methodology helps you stay structured even under pressure.

Unknown Systems and Dangerous Assumptions

You are often called out for assets you are not familiar with. This leads to assumptions about operating principles that may be incorrect, underestimation of energy levels such as pressure, voltage, or temperature, and improvisation with tools that are not suitable for the specific system. You also sometimes forget specific safety protocols that apply to that type of asset.
When dealing with unfamiliar systems, it is crucial to consult the documentation first. Ask an experienced colleague or the supplier for guidance if you are unsure. "Not knowing" is no disgrace, but acting incorrectly is. The time you invest in understanding the system will usually pay for itself many times over.

The Complexity of Energy Isolation

Modern installations often have multiple energy sources that you must secure. In addition to the obvious electrical main power supply, emergency power and UPS systems may be active. Pneumatic systems often have compressed air in various circuits, while hydraulic systems can be under high pressure. Don't forget mechanical energy either, such as spring tension, gravity, or rotational energy from rotating masses.
The major danger is that assets can "store" energy in capacitors, accumulators, springs, or rotating masses. Even after shutdown, these can be dangerous. Therefore, use the LOTOTO procedure

Working in Confined Spaces

Failures often occur in underground cable ducts, tanks and vessels, crawl spaces above ceilings, or inside machines with limited access. These locations carry specific risks such as oxygen deficiency or toxic gases, difficult evacuation in an emergency, entrapment by moving parts, and heat accumulation in confined spaces.

10.5. Cascading Failures — The Domino Effect

A failure can cause a domino effect where one problem triggers other failures. Failing cooling leads to overheating elsewhere in the system, pressure drops cause safety valves to fail, power outages disable alarm systems, and vibrating machines can damage connected pipelines.
The danger is that you become focused on solving the first failure, while new hazards are emerging in the meantime. Therefore, keep an overview of the entire system, not just the immediate failure. Communicate regularly with operators about the system status and stop when in doubt to reassess the situation.

The danger is that you become focused on solving the first failure, while new hazards arise in the meantime. Therefore, maintain an overview of the entire system, not just the immediate failure. Communicate regularly with operators about the system status, and when in doubt, stop to reassess the situation.

Night Shift and Reduced Alertness

Failures often occur at night or on weekends, when less support is available from colleagues and suppliers. Your alertness is reduced by fatigue, there is less light which makes it easier to make mistakes, and there is a tendency to "quickly" do something without full preparation.
Use extra lighting in these situations, more than you think you need. Take extra time for your LMRA, because fatigue increases the chance of errors. Don't hesitate to call a colleague, even in the middle of the night, and document everything with extra care, because details are forgotten more quickly when you are tired.

Frustration and the "It'll Probably Be Fine" Syndrome

After hours of searching for a cause, there is a tendency to address a "probable" cause without full verification. You skip safety steps with the thought "I've been doing this for years," change multiple things at once, or cut the LMRA short because you "are in a hurry."
Just after a long search, it is important to be extra disciplined. Frustration leads to mistakes, so take a short break if you notice you are becoming careless. The 5-step methodology helps you stay systematic in these situations as well.

Communication Problems During a Crisis

During major failures, many people are involved: production managers applying pressure, other technicians "helping out," operators trying different settings in the meantime, and management wanting continuous updates. Miscommunication about who is doing what can lead to dangerous situations when multiple people are working on the same system simultaneously.
Make clear work agreements about who does what, and schedule regular update moments instead of allowing constant interruptions. During critical work, one person must be responsible, while others assist according to clear agreements.

Recognition and Prevention

Be aware of these patterns. Experienced technicians recognize these situations and automatically take extra precautions. If you recognize any of the above situations, consciously slow down a step and double-check your safety measures.
Remember that a day later in the hospital is always slower than ten minutes of extra preparation. The systematic 5-step approach not only helps you find the root cause faster, but also keeps you safer, because you work in a structured way instead of acting impulsively under pressure.

11. Testing the Repaired System

As already mentioned, we will not go into the technical details of the repair, because those are different for every failure. We will therefore continue this story after you have completed your repair. To make sure the failure has now been resolved, it is necessary to test the system. Of course, it is always possible that something went wrong during the repair or that the actual cause of the problem has still not been resolved.

11.1. Phased Testing

The best sequence is to first test the repaired or replaced components separately. Do all motors rotate in the correct direction? Does the shaft rotate smoothly after the bearings have been replaced? Are all inputs and outputs correct?
If the components have been tested separately, it is possible to test the entire system one level higher. Perhaps a limit switch has shifted slightly somewhere, or a valve is still closed.
If possible, you should let the system run at least one complete cycle.
Only when everything has been tested and the system is working again can you be certain that the actual cause of the problem has been found and the repair was successful.

Together with the Operator

Always discuss with the operators of the asset whether there are other points they need to test or check. Maybe you repaired a leak and at this moment you cannot verify whether it has truly been fixed. The operator can check it for you. It would be considered unprofessional if you report a failure as repaired, while it turns out it is still failing.

11.2. If it doesn't work

If it turns out that the system is not yet functioning properly, you must start over and think: did something go wrong during the repair, or did we address the wrong root cause?

If you are convinced that you made no mistakes and forgot nothing, but the system still does not work, then the cause you thought was responsible for your problem was not the correct one. Go back to your list of potential causes and select the next one on your list.

Dealing Systematically with Failed Repairs

Not every repair succeeds on the first attempt, and that is not a sign of failure as a breakdown technician. Complex systems can have multiple causes or cascading problems in which the original failure has caused other problems that now also need to be resolved. The most important thing is that you keep working systematically instead of panicking.

With multiple causes, the problem often lies in the assumption that every failure has exactly one cause. In reality, a failure can result from a combination of factors. For example: a bearing begins to falter due to normal wear, causing increased vibrations, which in turn shake loose an electrical connection. You then need to both replace the bearing and repair the connection to get the system fully operational again.

Cascading problems are even more complex. The original failure can have caused a domino effect in which other components have been damaged or knocked out of their optimal operating condition. An overflowing tank, for example, can result from a fault in the cleaning cycle, but can also contaminate sensors and damage electrical connections. Resolving the original problem does not then automatically resolve all its consequences.

When your repair does not produce the desired result, it is crucial to go back to your list of possible causes.
Carefully document what you have tried and what the result was. This prevents you from going in circles by repeatedly trying the same approach. Record not only what you have replaced or adjusted, but also how the system responded afterward. This information is valuable for the next analysis cycle.

Know when to bring in outside help. If you have systematically applied the 5-step methodology, have ruled out all logical causes, and have performed the correct repairs without success, it is time to consult a specialist. A specialist can bring fresh expertise or have access to specific diagnostic tools that you do not have at your disposal.

It is important that you feel no shame about complex failures that are not resolved immediately. Some problems are inherently complex and require multiple analysis cycles to fully understand. The systematic approach in this book helps you manage this complexity, but does not always guarantee success on the first attempt. What is guaranteed is that a systematic approach will bring you to the final solution faster than chaotic trial-and-error work.

12. Failure Reporting

It is important that you write a good report on the failure. Do this shortly after completing the repair. Do not wait too long. There are various reasons to write a failure report.
In a number of cases, companies need the reports to correctly account for the costs and hours incurred. This is especially important when you work for customers who receive an invoice for the work you have performed.

Failure reports are also necessary to structurally improve assets. Your report will be read by people whose job it is to continuously improve assets. In some organizations these people are called reliability engineers or maintenance engineers. They regularly analyze which failures lead to the greatest losses for your organization, for example, because they occur most frequently, because they last the longest, or because your organization loses the most money on them. They also analyze which failures pose the greatest risks to safety, the environment, quality, or sustainability. They then develop modifications or preventive actions to ensure that the risk of the most significant failures is reduced to an acceptable level. They can only perform these analyses properly if they have the right information about failures that occur. They need your report.

Failure reports are also necessary for yourself, as a maintenance technician. They are useful when troubleshooting the same failure if it happens to occur again. Many failures cannot be eliminated through modifications or preventive maintenance. This means they will eventually recur. A good description of the current failure can then help you or your colleague resolve the failure more quickly. In order to be able to retrieve the failure, the title should reflect the most important symptom. Suppose a machine fails to start and the cause turns out to be an air pressure setting that is too low. Then the title should be 'machine fails to start' and not 'air pressure too low'. After all, the next time you encounter the failure, you will know that the machine is not starting — not that the air pressure is the cause.

You can also use the failure report to let the work planners in your maintenance department know that they still need to schedule some follow-up work as a result of this failure. Perhaps you replaced a broken wire with a temporary one that hasn't been routed cleanly yet, or you were unable to permanently repair a leak but temporarily placed a drip tray underneath it. In all such cases, the maintenance department still needs to take follow-up actions. In some companies, you would then need to create a new work request. Some other companies also use failure reports to communicate that follow-up actions are still required.

12.1. Minimum Content of a Failure Report

To achieve all of the above goals, a failure report must contain the following information:

- Title (With this title it appears in lists and reports.)
- Names of the parties involved (maintenance and production personnel)
- Date and time (start of failure, start of repair, end of repair)
- Which system had the failure? (asset identification)
- What were the symptoms? (all observed anomalies)
- What was the cause? (direct technical cause)

- What was the underlying cause? (use the 5-why method from section 9.1)

- What was done to resolve the failure? (repairs performed)
- Settings and parameters after correction
- Were there any other consequences? (secondary damage, safety risks)
- Are there any follow-up actions required? (permanent repair, preventive measures)
- Further details (specifics, learning actions)
- How many man-hours did you work on the failure? (including travel time)
- What materials did you use? (parts, consumables)
- How long was production down? (from failure to restart)
- Were there other consequences, such as product rejection? (quality loss, scrap)

12.2. Modern Reporting Systems

In many companies, you enter your failure report into an automated system. A well-designed system automatically fills in as much information as possible, such as which parts you retrieved from the warehouse. It is best if the system is set up so that all information you still need to enter manually is clearly requested in separate fields. If you have to enter everything in one large text field, you can use the above list as a memory aid.
Often, drawings, photographs, or sketches can be a great help in clarifying exactly what was going on. Many systems have apps for your phone for that purpose.

FAILURE REPORT		Date	Department	Asset name	ID No.
		31-10-2025	Drying	Brush machine	X234
Title	Brush seized and frame bent				
Technician(s)	MvdH, JvD		Operator(s)	JJ	
Failure start	Failure end	Production downtime	Repair start	Repair end	Man-hours worked
13.45	20.55	7.10	14.00	20.30	13
Impact	Safety [ ]	Environment [ ]	Quality [x]	Regulatory [ ]	
Symptoms	Alarm FC4555 came in at the control room. On site the brush was found to be jammed under the cleaner. The frame of the cleaner was severely bent. No anomalies had been noticed beforehand.				
Direct cause	Brush seized in reciprocating motion, cylinder rod was stuck				
Repair performed	Everything freed and cleaned. Frame straightened. Brush and cylinder replaced.				
Settings / parameters after correction	As before: 20 revolutions per minute and 3 cycles per minute				
5 x Why	Description		Materials	Quantity	No.
1	Brush seized		Brush	1	5123534
2	Cylinder rod seized		Cover	1	6345232
3	Severe contamination		Bearing	2	3534233
4	Protective cover torn		Cylinder	1	9944654
5	Too short and cannot be properly secured				
Follow-up actions required?	During scheduled shutdown: check all fastenings and realign frame. Install longer protective covers for cylinder rod with improved fixings.				
Remarks / lessons learned					

Figure 12-1 Example of a failure report

12.3. The Importance of Underlying Causes

Don't forget to include the underlying causes in your report as well. This is crucial for organizational learning and preventing recurrence.
Suppose that:

- → a machine crashes,
- → because the rod of an air cylinder gets stuck in a sliding bushing;
- → because the rod is contaminated with sticky product,
- → because the protective bellows around the rod is damaged...

Try to determine why the protective bellow failed:

- Is it too tight?
- Is it not flexible enough?
- Is it too stiff?
- Was it not correctly installed?
- ...

Your input can help the reliability engineer come up with a good solution. He cannot look inside the machine while it is running and would only see a properly functioning machine and an intact protective bellows. You were there. You had the protective bellows in your hands. You can better than anyone else assess why it broke.

13. The follow-up

Shortly after a failure, there should be a meeting with a number of people who each have their own expertise about the system. For major failures, this can be a separate meeting. For most minor failures, it is best if each production department holds its daily meeting, with someone from the departments responsible for maintenance, production, and other stakeholders.
The people in this meeting can discuss the possible causes of the failure and also discuss what they can do to prevent the problem from recurring. It is wise to document as many facts and ideas about the problem as possible, while all the data is still available and fresh in memory.

13.1. Realistic Prioritization

It is not possible to immediately prevent the recurrence of every failure: there will simply be too much work to handle. An investigation to get to the root of every problem, devise a solution, and implement it simply costs more resources than most organizations have. This usually leads to many unfinished studies. It is better to set clear priorities for the most important problems.
It is also not wise to permanently prevent all possible problems. Some problems occur so rarely and have such minor consequences that they are not worth the effort.

13.2. Correct Management Approach

In this daily meeting, the responsible manager should also ask those involved what they have done to perform the work safely. He could, for example, ask them about their Last Minute Risk Analysis. As described in chapter 10, LMRA is essential for safe failure work.
What he should never ask is 'why did it take so long?' This question is another way of saying 'hurry up.' It leads to many accidents every year. The focus should be on: "How did you work safely?" and "What can we learn for next time?"

13.3. Horizontal Thinking

Another thing to do the following day is to think horizontally. Thinking horizontally means you need to check whether this failure can also occur in other systems. In many cases, a number of systems are built the same way by the same people at the same time. If one system has a defect, this can of course also occur in those comparable systems. Perhaps a simple inspection can prevent many problems.

If a locking bolt has come loose on an agitator, it is necessary to also check all other locking bolts on similar agitators in your plant. This is essentially what product manufacturers do when they issue a recall for a specific product. All car brands that produce more than thirty thousand units per year have conducted at least six recalls between 2014 and 2024.

Note! By this we do not mean that the system must be completely dismantled, but rather that relatively simple checks should be performed to see what is going on. This also does not need to happen immediately. In general, the installation will first be restarted and the necessary inspections can be carried out during a scheduled shutdown.

13.4. Preventive Implementation

Horizontal thinking turns out to be a simple way to prevent failures. Moreover, your management will show little understanding if the same failure occurs twice with two different machines. They will then rightfully ask: 'you knew this could go wrong, why didn't you do anything about it?'

A good after-action meeting can turn a failure into a structural improvement that prevents future problems. Root Cause Analysis (RCA) teams can continue the systematic approach from the 5-step methodology to the organizational level.

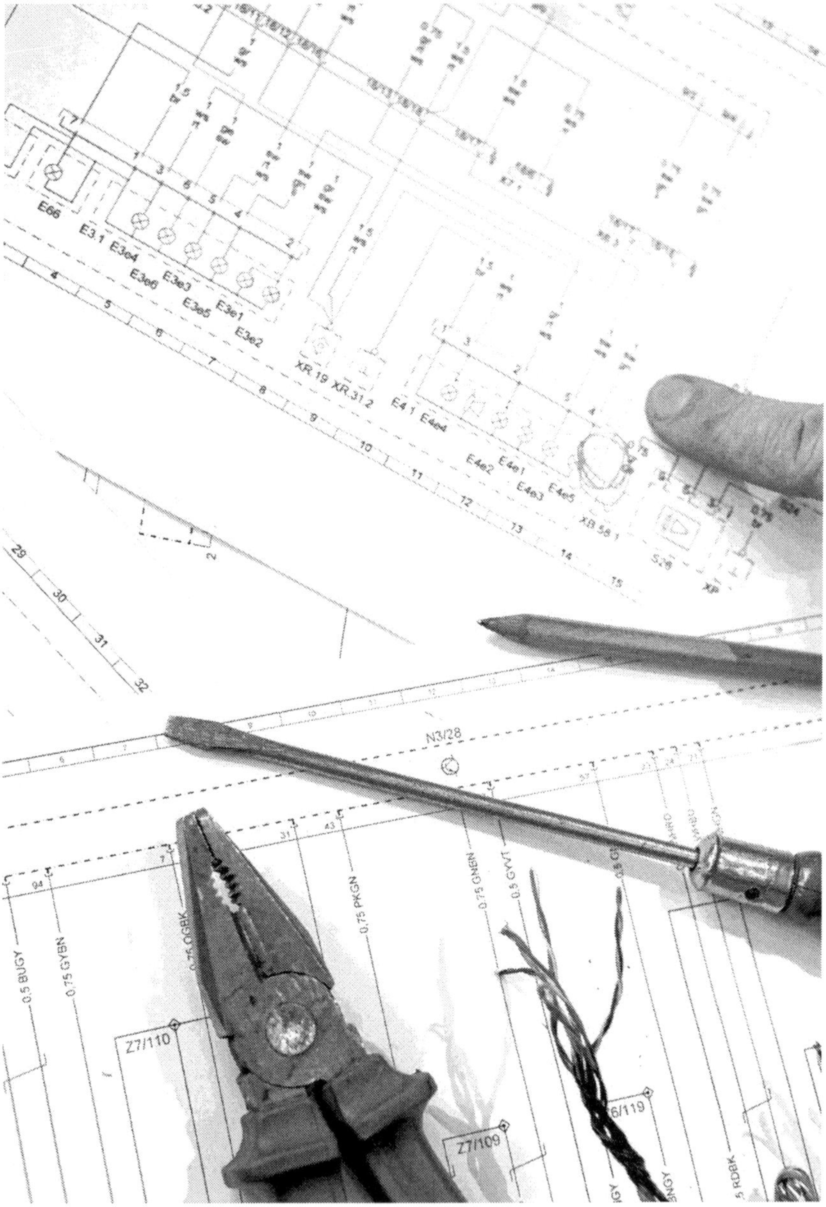
E66
E3.1
E3e4
E3e6
E3e3
E3e5
E3e1
E3e2
XR 19
XR 31 2
E4 1
E4e4
E4e2
E4e1
E4e3
E4e5
XB 58 1
N3/28
0,5 BUGY
0,75 GYBN
0,75 PKGN
0,75 GNBN
0,5 GYVT
Z7/110
Z7/109

14. Prepared for Future Failures

At the moment a failure occurs, it is too late to prepare yourself. Yet there is much you can do to prepare for failures in general.
In this chapter, you can read about a number of things that a maintenance technician can do himself, as well as a number of things that your company or organization can do.

14.1. What can a breakdown technician do?

To resolve a failure quickly and correctly the first time, it is necessary to know how your assets work.
Study how all the assets work that you should be able to repair. Read the manuals and make sure you truly understand what is in them. Take courses if possible. Many asset manufacturers offer such courses.

Look at the situation in practice as well. Take the time to carefully check the entire operation of the asset and observe all steps of the process. Which parts move? How fast? How far do they move? Which lights are on? Which messages appear on the screen? Direct observation is the only way to truly get to know a machine or other asset. It may sound strange, but if you simply spend a morning just watching a running machine, you will easily earn back that time later, because you can resolve failures on that machine much faster.
Take photos and notes. In every process there are many meters and sensors for, for example, temperature, pressure, or current. Record their values during normal operation and possibly under special circumstances. These baseline values are worth their weight in gold during failure diagnosis.

Talk to operators and more experienced colleagues. Ask them what the strengths and weaknesses of the asset are and what problems they have experienced with it. Their knowledge of subtle signals can help you recognize failures early.

Personal Toolbox

Make sure you have the right tools to properly investigate a failure. You can think of simple tools, such as:

- A flashlight - often essential in dark spaces
- Pen and paper - for quick sketches and notes
- Measuring tape - for dimensions and positioning
- Smartphone with the apps from chapter 7

Proactive approach

If you work on failure response duties within a specific production department, make sure you know what is going on. What products are they producing this week? Are there any problems, changes, or maintenance activities?

If you can be called at home outside of working hours, it is very useful to spend the last hour before going home making a round through the department and asking operators whether there are any problems. In a number of cases, you'll already hear about small problems that you can resolve on the spot, before they grow into major problems that require you to leave home.

14.2. What can the organization do?

Naturally, the organization must support the failure response technicians in acquiring the knowledge and resources from the previous section.

Documentation and Accessibility

In addition, the organization must ensure that all knowledge and documentation of the assets are up-to-date and easily accessible. Think of drawings, diagrams, manuals, and similar items. Enterprise Asset Management software can be very useful for the structured digital storage of this information.

Furthermore, the organization can ensure that the failure reports for every failure are clear and prepared in accordance with the guidelines from chapter 12.

Modern Technology and Systems

Another thing the organization can do is to make sure that every machine has clear alarms. It doesn't have to cost much to modify a machine's control software so that a screen clearly indicates which part of a machine is causing a problem and where it is located.

In a number of cases, a manufacturer of equipment can help resolve failures remotely. This is simplified when the supplier has direct remote access to automated systems. Remote expert support via video call can be valuable, where a specialist assists remotely without being physically present.

Practical Provisions

Also make a few simple overview drawings to indicate where all the sensors and motors are located. It often happens that a maintenance technician has found in the control system of a machine which sensor or motor is causing the problem, but cannot locate it on the actual machine. Make sure that all parts of assets have proper labels or tags. QR codes can provide direct access to digital documentation and failure history.

Spare Parts and Tools

Make sure you have the right spare parts. A thorough analysis of spare parts requirements often pays for itself after just one or two failures. You can read more about this in book 6 of this series.

Sometimes it is useful to install components redundantly, so that operators can immediately switch to the other component in the event of one component failing.

A company can also enter into contracts with suppliers so that they deliver parts 24 hours a day, 7 days a week.

Consider the availability of the right tools and equipment. To read out failures on many assets, you need a laptop with the right software. Make sure that laptop with software is available and teach people how to use it.

Accessibility and Safety

Make sure the asset and all its components are easily accessible. Do not place them flush against a wall. Do not run pipes or cables over parts that may need to be replaced. Many plants have been completely shut down because an air hose had to be removed to replace a component. If there is no valve to isolate a small section of your air network, the entire plant has to be shut down. So also make sure there are sufficient local valves and switches in all lines for air, electricity, water, steam, and the like.

External expertise

Your own maintenance technicians can't solve everything. Make sure you have good contracts with suppliers and manufacturers, so they can be on-site quickly in the event of a failure. If necessary, it is sufficient for them to observe while one of your own employees works with a virtual reality headset. A specialist can then watch along from a different location and provide instructions.

Epilogue: Failure Analysis in Perspective

James, Eighteen Months Later

James walked through the machine hall of ServiceTech. Eighteen months earlier he had been here during that disastrous night at FreshFood Industries — twelve hours of chaos over a loose plug. Now his phone showed a different story: average resolution time down from 4.2 to 2.1 hours per failure, 80% fewer unnecessary parts replacements.

"Nine times out of ten, I find the root cause on the first try now," he said. "And more importantly: often I know exactly why it's that cause, before I even loosen a single bolt."

But the eighteen months had also brought new insights. Failures that kept recurring despite correct repairs. Parts that were not available in time. Systems that had failed due to design flaws that no maintenance technician could resolve.

"Systematically solving failures has transformed us," James admitted. "But I also saw where it stops."

What Failure Analysis Does and Does Not Solve

James's transformation demonstrates the power of systematic failure analysis. You now know how to find the root cause using the 5-step methodology, how to work safely under pressure, and why asking open-ended questions is more important than immediately turning wrenches.

But as James discovered, challenges remain:

Recurring failures: Some problems keep coming back because the underlying cause runs deeper — a design flaw, incorrect specification, or structural process problem that a single repair cannot resolve.

Prevention instead of reaction: You are now excellent at resolving failures, but wouldn't it be better if they didn't occur at all? Systematic preventive maintenance prevents many of the failures you have now become so skilled at solving.

Inventory management: You can now quickly find which part has failed, but if that part has a four-week lead time while production is standing still...

Design limitations: Even the best breakdown technician cannot solve fundamental design flaws. Some assets are simply poorly designed or incorrectly specified for their application.

Different Routes to the Same Goal

James started, as many breakdown technicians do, by improving his intervention skills. It quickly delivered measurable results and reduced the stress of acute situations. But it is not the only approach.

Some organizations start by systematically reducing failures — through root cause analysis of recurring problems. This works especially well when a few major failures are responsible for most of the losses.

Others start with preventive maintenance because they want to prevent failures from occurring in the first place. Or with work flow control because their maintenance department reacts chaotically to every failure without prioritization.

The Nine Disciplines for Structurally Reliable Assets

The challenges that remain — from work organization to preventive maintenance, from inventory management to design flaws — each call for their own systematic approach. This book series covers nine disciplines that together answer the question: "What must an organization do to make its technical assets perform as well as possible?"

1. Maintenance Management – The Overarching Framework
2. Planning and Scheduling in Maintenance– Efficient Organization of Work Activities
3. Troubleshooting Breakdowns– Effective Intervention in Acute Problems
4. Root Cause Analysis in Maintenance– Addressing Recurring Problems
5. Preventive Maintenance – Proactive Asset Care
6. Spare Parts Management – Optimal Inventory Control
7. Precision Maintenance– Flawless Installation and Overhaul
8. Engineering for Reliability – Reliability During Design
9. Autonomous Maintenance – Operator Involvement

The nine books follow a logical structure in which each book builds on knowledge from earlier parts. The implementation order in your organization may differ, however — start where the pain is greatest and where you can achieve results quickly. In the introduction of this book you will find a complete overview of all parts and how they relate to one another.

More information about the complete series
For a current overview of all books, additional information and ordering options, visit:

vadeo.nl/books.html

Every discipline you add makes the others more effective. They reinforce each other. Good work flow control ensures you have time for precision work. Preventive maintenance works better when operators detect deviations early. Reliability engineering prevents problems that you would otherwise have to address structurally. The order in which you implement them depends on what your organization needs most.

Your Next Step

You have now learned a methodology that works. ServiceTech's results — and the thousands of field technicians who apply similar approaches — prove that the 5-step methodology transforms failure resolution.

The question is: do you stop here, or do you build further?
For now: apply what you have learned in this book.

Use the 5-step methodology. Ask the 7WH questions. Eliminate causes systematically. Measure your results. Celebrate your successes.

James's advice: "Start where the pain is. For us, that was the chaos during failures. The systematic approach made an immediate difference — less stress, faster results, safer work. But later we discovered that other aspects were holding us back. Root cause analysis helped us permanently solve recurring problems. Preventive maintenance prevents many failures from occurring in the first place. It's not about the order, but about the systematic approach."

Appendix A: Template for failure report

FAILURE REPORT		Date	Department	Asset name	ID No.
Title					
Technician(s)			Operators		
Failure start	Failure end	Production downtime	Repair start	Repair end	Hours worked
Impact	Safety [ ]	Environment [ ]	Quality [ ]	Regulatory [ ]	
Symptoms					
Direct cause					
Repair performed					
Settings / parameters after correction					
5 x Why 1			Materials description	Quantity	No.
2					
3					
4					
5					
Follow-up actions required?					
Remarks / lessons learned					

Appendix B: Guided Worksheet

This form helps you to map out all relevant information when solving a failure. It is not a failure report.

Form for Targeted Failure Analysis

Name:	
Datum:	
Machine	

Information about the failure

Description of symptom

		Source (*) 1	2	3
What	**What NOT**			
When	**When NOT**			
Where	**Where NOT**			
Which	**Which NOT**			
How	**How NOT**			

(*) 1 = heard, 2 = seen, 3 = measured

Possible causes

	Description	Eliminated, because	No.
1			
2			
3			
4			
5			
6			
7			
8			
9			
10			
11			
12			

Probable cause and underlying causes

Risks and points of attention during repair

Risks	Measures

Tests

Functional test: from small to large

Appendix C: Questionnaire

This list provides suggested questions to help map out the problem. It is very important to also establish what the problem is NOT. If the problem does not occur at a certain time, in a certain way, or at a certain location, you can rule out a number of possible causes.

WHAT?

What is the problem? • What exactly is wrong? • What is the asset doing incorrectly? • What deviation does the product show? • What happened when the failure occurred? • What did people see happen? • What did they hear? • What was the operator doing just before the failure? • What was done after the first symptoms appeared? • What is the normal behavior of the installation?

WHY?

Why is this failure a problem? • Why do you think the asset is not functioning correctly?

HOW?

How significant is the failure? • How serious is the deviation? • How do you detect the failure?

WHERE?

Where exactly is the deviation/damage located? • Where can the failure be seen/heard/felt? • Where in the machine? • Where on the product?

WHEN?

When does the failure occur — constantly or intermittently? • When did the failure first occur? • Does the failure occur constantly or intermittently? • Did the failure appear suddenly or worsen gradually?

WHICH?

Which asset exactly has the failure? • Which products show the deviation? • Which settings are being used?

WHO?

Who first noticed the failure? • Who worked on the asset before the failure? • Who can provide more information?

CHANGES?

Has anything changed recently on the asset? • Are there changes in:

- Settings
- Raw materials
- Environmental conditions
- Operating personnel
- Maintenance activities
- Software/controls
- Procedures

People often do not consider something a change, while it is one as far as the asset is concerned. Therefore, ask about all activities on or near the asset recently. Has the asset been:

- Serviced
- Converted
- Cleaned
- Improved
- Repaired
- Adjusted

DIFFERENCES?

What is the difference between products/assets that work correctly and those that do not? • What circumstances are different when the failure does occur compared to when it does not?

Tip: Assets rarely fail "out of nowhere" — there is almost always an event or change that triggers the failure. Be aware of recent changes!

Figures

Glossary — Terminology and Acronyms

5M Methodology A systematic thinking model for root cause analysis with five categories: Man (operating errors, assembly errors), Material (wrong raw materials, components), Method (procedures, instructions), Medium (environmental conditions), Machine (technical causes such as wear, corrosion).

5-Step Methodology A systematic approach to failure resolution: 1) Describe the failure, 2) Establish a list of possible causes, 3) Eliminate causes, 4) Repair, 5) Follow-up.

5-Why Method A technique that asks 'why?' five times to move from symptom to root cause; effective for problems with a linear causal chain.

7WH Method A systematic questioning method for failure analysis: What, Why, How, Where, When, Which, Who. Fundamental principle: it is always more important to know where a problem does NOT occur than where it DOES occur.

Asset Knowledge Understanding of how an asset operates under normal conditions. Baseline values of gauges, normal sounds, movement patterns. Fundamental for rapid failure diagnosis.

Augmented Reality (AR) Technology for remote support where a specialist watches via video call and can place digital annotations on the screen of the local technician.

Baseline Information Documented normal values during operation: temperatures, pressures, current levels, rotation speeds. Invaluable as a reference during failure diagnosis.

Cascading Failures A domino effect in which one failure triggers other failures. Danger: focusing on the first failure while new hazards develop simultaneously.

Changes Events or modifications that trigger a failure. The vast majority of failures occur as a result of a change. Use various words: improved, optimized, adjusted, modified, replaced, maintained, accelerated.

Closed-ended question A yes/no question for verification after open-ended questions have been asked. Apply only when you already have certain ideas about possible causes.

Confined Space A location with specific risks such as underground cable gutters, tanks, and crawl spaces. Requires atmosphere measurement, constant communication with an outside post, a harness and rescue line, and adequate ventilation.

Confirmation Bias A cognitive error in which all observations are interpreted to confirm the original theory, instead of objectively considering other possibilities.

Core Question of Failure Analysis The central question in root cause elimination: 'Does this cause explain ALL symptoms that occur?' A potential cause must explain every observed symptom, time pattern, and location.

Definitive Solution A permanent repair that completely eliminates the root cause. Preferable over a temporary solution as soon as circumstances allow.

Differences Systematically searching for what is different between situations where a failure does and does not occur. Fundamental to the elimination of causes.

EAM (Enterprise Asset Management) Software for planning, execution, and registration of maintenance and spare parts management.

Energy Sources All forms of energy in systems that can pose a hazard: electrical, mechanical (spring tension, gravity, rotation), pneumatic, hydraulic, thermal, chemical. Even after shutdown, these systems can store energy.

Failure Any situation in which an asset does not meet the specified requirements. Can range from complete breakdown to deviations in noise, leakage, or performance. Also referred to as a defect or fault.

Failure History Historical documentation of previous failures of an asset. A large portion of failures occur more than once — good documentation saves future diagnostic time.

Failure Mode The specific way in which a component or system can fail to perform its function.

Failure Pattern The characteristic development of the probability of failure over time; six different patterns (A–F) determine which maintenance strategy is most appropriate.

Failure Report Documentation of a failure with at a minimum: personnel involved, date/time, asset identification, symptoms, cause, root cause, repairs performed, secondary damage, follow-up actions, man-hours, materials used, production downtime.

Horizontal Implementation Applying a solution preventively to comparable assets before a failure occurs there. Essential for structural improvement.

Horizontal Thinking Checking whether the same failure can also occur in other comparable assets in the organization. Preventive inspection prevents recurrence elsewhere.

Leading Question A question in which an answer is already suggested. Avoid these at the start of an analysis as they lead to incorrect answers.

LMRA (Last Minute Risk Analysis) A 3-step safety check for every maintenance task: 1) Identify hazards, 2) Determine measures, 3) Implement measures before you start. Essential for failure work, where 75% of industrial accidents occur.

LOTOTO (Lock Out, Tag Out, Try Out) A safety procedure for isolating energy sources before maintenance begins: locking out, tagging, and testing the isolation.

Maintenance Technician A person who resolves a failure, regardless of job title. The logic of the systematic approach applies equally to mechanical, electrical, electronic, chemical, or software systems.

Open-ended question A question that begins with a question word (what, how, which, where, why, who) and gives no direction to the answer. Essential at the beginning of an analysis to avoid putting words in someone's mouth.

Phased Testing A systematic test method: first repaired parts separately, then the system one level higher, finally the complete cycle. Prevents discovery of failures after a complete restart.

Predictive Analytics AI systems that recognize patterns in sensor data indicating impending failures. Alerts for preventive action before an actual failure occurs.

QR Codes A digital link on an asset for direct access to documentation, failure history, maintenance reports, and technical specifications via smartphone.

Remote Expert Support Video calling with a specialist who monitors and advises in real time, without being physically present. The local technician remains responsible for safety and physical actions.

Resolution Time Time from failure notification to a repaired and tested system. A systematic approach reduces this by an average of 50% compared to trial-and-error.

Smartphone Diagnostics Use of smartphone sensors for failure analysis: camera for slow-motion vibration analysis, microphone for sound analysis, GPS for location documentation.

Symptom An observable deviation from normal behavior of an asset. All symptoms together form the failure pattern that points to the cause.

Temporary Solution A solution that resolves a failure in the short term but not permanently. Permitted when the correct parts are not available, the repair takes too long, or as an alternative to managing consequences. Requires clear documentation for definitive follow-up.

Trial-and-Error An unstructured approach in which components are replaced without systematic analysis. Leads to longer resolution times, unnecessary costs, and potentially new damage.

Tunnel Vision A dangerous pitfall in which a maintenance technician clings to one preferred theory and becomes blind to other possibilities. Occurs especially under time pressure.

Underlying Cause The cause behind the cause. A broken bearing can be the result of a faulty lubrication system, which in turn is the result of an empty grease reservoir, which in turn is the result of having no procedure. Analyze at a minimum of five levels deep.

Verifying Information Checking whether received information is correct. People perceive things incorrectly, interpret things differently, unconsciously fill in gaps, and may want to keep things quiet.

Vertical Thinking Searching for underlying causes (why did this happen?) and further consequences (what do I still need to check before starting up?).

Acronyms and Abbreviations

CMMS Computerized Maintenance Management System

EAM Enterprise Asset Management

FAT Factory Acceptance Test

LMRA Last Minute Risk Analysis

LOTOTO Lock Out, Tag Out, Try Out

MTBF Mean Time Between Failures

MTTR Mean Time To Repair

OPL One Point Lesson

P&ID Piping and Instrumentation Diagram

PPE Personal Protective Equipment

RCA Root Cause Analysis

SCADA Supervisory Control and Data Acquisition

SMART Specific, Measurable, Achievable, Relevant, Time-bound

TPM Total Productive Maintenance

Made in United States
Orlando, FL
04 November 2026

90209063R00074